HIGH SKIES
AND YELLOW RAIN

ADRIAN BERRY

𝔇𝔞𝔦𝔩𝔶 𝔗𝔢𝔩𝔢𝔤𝔯𝔞𝔭𝔥

Published by the Daily Telegraph
135 Fleet Street, London EC4P 4BL

First published 1983
© Adrian Berry 1983
ISBN 0 86367 020 2

Typeset by Design Perspective Ltd, London N10

Printed by Cox and Wyman Ltd, Reading

Contents

HIGH SKIES
AND YELLOW RAIN

By the same author
The Next Ten Thousand Years
The Iron Sun
From Apes to Astronauts
The Super-Intelligent Machine

Fiction
Koyama's Diamond

ERRATA

Owing to production difficulties it is regretted that three typographical errors appear:

On page 81 the name of Douglas Hofstadter has been printed twice.

On page 136 there should be the footnote, *Facts and Fallacies: A Book of Definitive Mistakes and Fallacies* by Chris Morgan and David Langford (Webb and Bower).

On page 142 line 22, ancestors should read descendants.

FOREWORD

WHEN Adrian Berry invited me to write a foreword to his new book, I was delighted to accept. He is one of the select group of writers who possess the enviable facility of presenting modern science and technology to the readers of the daily newspapers. I know of few more difficult tasks for a writer. It is easy to write about science and all too easy to write nonsense. Every profess-ional scientist knows how hard it is to write accurately in everyday language about his own subject. Readers of the *Daily Telegraph* are well aware that they can turn to Adrian Berry's articles in the confidence that they will find accurate and concise accounts of a wide range of scientific and technological topics.

The range of Adrian Berry's understanding of the modern scientific world is evident in this collection of articles. The astronomical section takes us to the forefront of modern investi-gations. By analogy, the immensity of the universe is reduced to comprehensible terms. The telescopes in use today may be penetrating so far into space and so far back in time that we believe we have an understanding of the universe as it was billions of years ago – and that may be the epoch when space and time began. Are we the only forms of life in this cosmos? That is a speculative question and here you can discover how the scientists of today react to this query.

In the autumn of 1957 the Soviets startled the world by launching the world's first artificial earth satellite – the Sputnik. The launching rocket was developed as an intercontinental ballistic missile and I am glad that Adrian Berry writes so sensibly about the contemporary military associations of space launchings. Tremendous possibilities for good or evil are narrowly entwined in the modern world of space activities. An important service is performed by anyone who places the military, commercial and scientific perspective of space before the public and here it is done admirably.

Although the mathematical principles on which computers are based may have been known for a considerable time, the techniques which have led to our modern computers are almost wholly a post-World War II development. There can be few examples in history where scientific devices have emerged from the laboratory to penetrate our normal life so dramatically. Will future developments lead to the creation of an 'artificial intelligence'? The limitations and future possibilities are well treated by Adrian Berry. So too are the present-day and future prospects of our technology – from satellite and cable television to the future super-high-speed train.

In one group of articles the charlatans of the scientific world are brilliantly exposed. The juxtaposition of the maniac and pretentious with the serious business of the scientific world illuminates the need for honesty and reliability in science writing. We are too close to the dramatic developments in scientific research and the increasing rapidity of the technological change to assess whether twentieth-century science will emerge as beneficial or disastrous for the human race. The nature of fundamental scientific research, and especially those parts of the activity chosen for technological development, are intimately entwined with the society in which they are pursued. In the initial act – be it in the study of the universe or in the fundamental mathematics of the computer – the possibilities for ultimate good or evil are present. In the end it is the people who make the choice and the sensible choice cannot be made in ignorance. We owe a great debt to Adrian Berry for helping so many of us to an understanding of these fundamental issues.

Sir Bernard Lovell O.B.E., LL.D., DS.c., F.R.S.

ASTRONOMY

THE age and vastness of the universe involve such colossal numbers that many people switch off their minds when confronted by them. But there is no reason to panic at these statistics. Space and time are finite. If the Sun were miraculously shrunk from its present diameter of nearly a million miles to that of an atomic proton, a 25th of a trillionth of an inch, the edge of the known universe, instead of being countless billions of light-years distant, would be a mere 300 feet away. And if a billion years were compressed into a 24-hour day, then Christ would have been born a fifth of a second ago, and World War II would have ended .003 seconds ago.

When contemplating these immensities, one is drawn to two conclusions: that the future of our descendants is virtually unlimited, but that this statement is only true if we discover, while there is still time, whether our Milky Way Galaxy is free of dangerous alien intelligence.

High skies

MERE adjectives cannot accurately suggest the size and timescales of the universe. Phrases like 'boundless vastness' and 'immeasurable eons' give an unsatisfactory picture because the universe, although very large, has finite boundaries. The latest estimate of the distance to those boundaries – the answer to the question, 'How high is the sky ?' – is about 120,000 million million million miles. For a light ray or a radio message to reach us from the edge of the unknown universe would take about 20,000 million years; for the speed of light, the greatest velocity that any object or message can travel, is a 'mere' 670 million mph.

Our Milky Way Galaxy, which contains about 150,000 million suns, is but part of a cosmos of some 100,000 million other galaxies. The number of suns in the universe, therefore, is the product of these numbers, which is 15,000 million million million stars, or, as astronomers would write it, 15×10^{21} (which is simply 15 followed by 21 zeros). The best way to appreciate the distances of this universe of some 10^{70} cubic miles is the 'scale down' method, in which the Sun is progressively reduced in size, and distances become shorter in proportion.

Although in reality nearly a million miles in diameter, imagine the Sun only one foot across. If it was situated in the centre of Piccadilly Circus, Earth would be a tiny ball 107 feet away and one tenth of an inch in diameter. Jupiter would be 560 feet away (1.2 inches in diameter) and Pluto, the furthest of the planets, around Hyde Park Corner, just under a mile distant.

So far, the foot-scale model of the universe is comfortably confined to central London. But with the inclusion of the stars things get out of control. With the Sun only a foot wide, a light year is reduced to 1,300 miles, and the nearest star, Proxima Centauri, is four and a third light years from us. So Proxima Centauri is in Kansas City.

Since our galaxy alone would extend on the foot-scale out into space beyond the orbit of Jupiter, 12 inches is clearly too large a size for the Sun. Let us therefore reduce the Sun to a bacterium, about one eight thousandth of an inch across. In this bacteria-scale Pluto's orbit, which previously bisected Hyde Park Corner, is now only half an inch across. Now a light year is a mere 60 feet, and the nearest star is 260 feet from the tiny Sun-bacterium in the centre of Piccadilly Circus.

This looks much more promising. We now have 51 stars within 1,000 feet of the statue of Eros, separated from each other by an average distance of 530 feet. Some famous stars are still inconveniently far away – Betelgeuse is three and a half miles, and Rigel is seven miles – but at least they are within bicycling distance. It is therefore a shock to realise that our galaxy alone extends from London to Naples, and the edge of the known universe is nearly twice as far away as the real Sun.

Let us now reduce the model Sun still further, to the size of a hydrogen atom, a 250 millionth of an inch in diameter. Now, surely, it will be possible to contain the universe in the Piccadilly area. At first sight it looks hopeful. Our galaxy, which formerly straddled London and Naples, is now only 225 feet across, and the Sun is 80 feet from the galactic centre. The entire galaxy will rest comfortably on an acre of ground.

But look for the other galaxies – and the trouble starts again. The great Andromeda galaxy, one of our nearest neighbours, is at Hyde Park Corner. The exploding galaxy in the Great Bear, M82, is in Hammersmith. A cluster of galaxies in Virgo is in Uxbridge. A giant cluster in Coma Berenice is in Stratford-on-Avon. The nearest quasar star is in Algiers, and the edge of the known universe is at Johannesburg.

To bring the entire universe into Piccadilly Circus, we must take one final step and reduce the Sun to the size of an atomic proton – with a diameter of a 25 million millionth of an inch. At last, the whole cosmos can be seen from the entrance of the London Pavilion cinema. Our galaxy is now 0.03 inches across, the nearest quasar star is 40 feet from it, and the edge of the known universe is just 300 feet away.

To reduce the universe into a locket, which could be worn round a man's neck on a chain, we would have to reduce the entire galaxy to the size of a 'quark', the smallest known atomic particle, which is about 0.000000000000003 of an inch across. Even smaller objects than the quark have been imagined. Scientists at Princeton have calculated the size of the 'ultimate unit', a particle of roughly one millionth of a thousand millionth of a thousand millionth of a thousand millionth of a centimetre in length. These scales also arc typical measurements of the universe, a place where objects are at once so large and so small that the human mind loses all power to contemplate them.

The time scales of the universe, like its distances, require just as radical a compression before they become easy to visualise.

Pretend that a thousand million years equals one 24-hour day. On this scale, the universe created itself with a cataclysmic explosion three weeks ago. About five days ago, the Sun was formed, and the planets took shape a couple of hours later. The first man-like creatures appeared just over eight hours ago. Christ was born a fifth of a second ago, and Hitler committed suicide .003 seconds ago.

The Sun is already showing the first signs of fatal decay. But it should burn at its present strength for the next five days. After that, it will start to expand into a red giant, turning Earth's oceans into clouds of boiling steam. Seven days from now, it will shrink to smaller than its previous size and Earth will become a frozen lump of rock.

What prospects await the human race in this immensity of time and space? In the very long-term, mankind has nothing to fear even from nuclear wars, or almost any other kind of disaster. It is improbable that the most catastrophic nuclear war would not leave at least a few thousand survivors. From these survivors, hardy evolution and five million years would soon produce a second, more brilliant race.

For what is five million years compared with five thousand million of the Sun's healthy future? Man might face permanent ruin if half, or even a third, of his tenure on this planet was threatened by some catastrophe, but to lose a thousandth need not worry him. To be pessimistic about the long-term future because of possible nuclear wars is as neurotic as to fear that 'next year will be miserable' because hangovers lasting nine hours will occasionally have to be endured.

But five thousand million years is the minimum that man has at his disposal. The Sun will die at the end of that time, but other stars will by then be starting their cycles. Many of them will doubtless have agreeable planets on which our descendants will settle.

'We are living at the very beginning of time,' wrote the great astronomer Sir James Jeans. 'We have come into being in the fresh glory of the dawn. Our descendants of far-off ages, looking down this long vista of time from the other end, will see our present age as the misty morning of the world's history. Our contemporaries of today will appear as dim heroic figures who fought their way through jungles of ignorance, error and superstition to discover truth.'

Powers of ten

A young man and a girl lie asleep in the sun after a picnic lunch somewhere in Chicago. A camera watches them from above. Suddenly, an extraordinary thing happens. The camera lens widens by a factor of 10. Now these two people no longer fill the field of view. They occupy only the central square in a large expanse of lawn.

Again, the view widens 10 times. Now the picnickers can barely be seen for the lawn they have chosen for the lunch appears as a mere strip of grass between a motorway and a harbour. Once more the lens widens tenfold. Lawn, motorway and harbour all now seem insignificant in the huge complexity of the city.

These progressive tenfold widenings are the theme of a magnificent illustrated book* by Philip and Phyllis Morrison, aimed at showing the scale of things in the universe, the very large and the very small.

The picture had originally been of an area one yard across. At the fifth widening it is 60 miles, showing most of Illinois on the

* Powers of Ten, *Philip and Phyllis Morrison (Scientific American Library, distributed by W.H. Freeman and Co., San Francisco)*

curved southern tip of Lake Erie. From now on, with each tenfold decrease in scale, we seem to move away from the earth into space at ever increasing speed. Let us examine the views produced by these tremendous accelerations.

At 600 miles we see much of the northern hemisphere, and at 6,000, the entire earth. At 60,000 miles, the earth is a small globe amidst a starry field, and at 600,000 it is all but invisible, ringed by the elliptical orbit of the moon.

On a scale of six million miles, the orbital paths of the two worlds are entwined in a broad swathe with a curve that is barely detectable. But when the scale is decreased to 60 million miles, this shared orbital path is but one of several that mark the planetary movements. Two more decreases of scale reveal the whole solar system, and with another, the picture is one light-year across. The positions of the planets can no longer be shown, and the sun is no more than a bright star.

Between 10 and a thousand light-years, the picture hardly seems to alter. The sun has long ago vanished in the seemingly boundless reaches of the Milky Way. Then, between 10,000 and 100,000 light-years, an awesome new change becomes apparent.

Our galaxy itself shrinks to a single object. At 10 million light-years it is one blob amid a dozen others. Still further away we rush. First we see our local group of galaxies, the Virgo Cluster; then, at 1,000 million light-years, space is almost void of matter. There are little bright smudges here and there, isolated from each other in an immense darkness. These are groups of clusters of galaxies, universes of universes in an almost empty universe.

Let us take a journey in the opposite direction of scale. A young man and a girl lie asleep in the sun ... The camera lens above them narrows by a factor of 10. At 10 centimetres, we see wrinkles on the man's hand. At a tenth of a millimetre we are truly in the sub-world of microscopy. At 10 microns (a micron is a millionth of a metre) we encounter a capillary vessel of the blood, and at a tenth of a micron we discover coils of DNA.

Down, down we go, into a world even smaller. Clusters of molecules give way to single molecules – just as galaxies gave way to clusters of galaxies. We are soon reminded that there are more atoms in that patch of human skin than there are people in the world. Clouds of electrons look like fields of stars on the microscale. At last, as a millionth of a micron, we reach the nucleus of a carbon atom. Deep inside the nucleus lie quarks, the smallest known particles, the very bedrock of matter.

Here are 42 steps, 42 powers of 10, taking us from the world of the gigantic to that of the minuscule, each one created by adding or removing a zero from the dimensions of the last. And of these 42 different views of the universe, it is remarkable that only about three are familiar to our everyday senses.

Time's winged chariot may be late arriving

WILL time come to an end, or will it go on for ever? The answer to this strange-seeming question that now so perplexes astronomers will, when definitely answered, more clearly indicate the ultimate destiny of thinking beings.

But what does the question mean? At first sight it appears to have no meaning, since 'time' is generally regarded as an abstract thing, which cannot be said to 'end' or 'begin'.

Yet this is not the case. All our present knowledge of the universe rests on the premise that there was a single beginning of everything, that matter, time and space simultanously came into existence in an explosion called the Big Bang, some 20,000 million years ago

But what, it is often asked, happened before the Big Bang? The question is meaningless. There was no 'before' since it marked the beginning of time. And again, it is asked, where did the Big Bang happen? But again, the question has no meaning, for the Big Bang created all the space that now exists. In short, it happened everywhere.

What had a beginning may surely have an end. The eloquent American astronomer Carl Sagan, in a recently published wide-ranging book,* surveys the night sky and presents the two opposing predictions of the future of time.

As far as we now understand that matter, there are only two possibilities. If the total weight, or 'mass', of the universe exceeds a certain critical limit, then all the galaxies (giant groups of stars) which are still rushing apart from each other under the impetus of the Big Bang, will eventually halt their expansion and collapse into a single object that will rapidly shrink into a state of infinitesimal size and infinite density, while retaining all its original mass. This event, if it occurs, will happen in about 50,000 million years.

If, on the other hand, the mass of the universe is below that critical limit – and there is as yet no consensus on the question – then the galaxies will continue to rush apart from each other for eternity. Life and energy will die, but of space and time there will be no end.

Which would we *prefer* to live in, a 'closed' collapsing universe, or an 'open', ever-expanding one that eventually 'dies' for all practical purposes from the exhaustion of nuclear fuel?

Many scientists, fascinated by the mysterious properties of the Big Bang, predict that a collapse would produce fresh Big Bangs, perhaps infinite in number, as time, space and matter, and ultimately, perhaps, civilisations themselves, are continuously re-created by these explosive invasions from another dimension.

An oscillating or a 'breathing' universe of this kind has its attractions, particularly to those of an apocalyptic disposition, who feel that a universe so filled with trouble and sorrow would surely be better the next time round.

But there may be no 'next time round'. Perhaps our Big Bang is the only one that will ever occur; in which case, to hope on these grounds for a collapsing universe is a dangerous dream. Let us look at the other possibility and see if it is preferable.

Consider that figure of 'about 50,000 million years' between now and the possible collapse. It seems an enormous amount of

* Cosmos, *Carl Sagan (Macdonald Futura, 1981)*

time, almost beyond human comprehension. But it is nothing compared with the number that is to follow! The number of billionths of a second in 50 billion years is but a tiny, imperceptible fraction of Freeman Dyson's Number, the number of years in which, according to the calculation of Professor Freeman J. Dyson, of the Institute for Advanced Study at Princeton, intelligent life will be possible in an ever-expanding universe whose mass is below the critical limit that would bring about gravitational collapse.

Let us be thankful that we were never required to learn Freeman Dyson's Number at school. For if very large numbers are held to be repellent, as many people so hold them, then it is a truly horrible statistic! Before all the nuclear fuel in the universe is exhausted, before all the stars have died – and before the deaths of all the shining stars that have yet to shine, through countless billions of generations of shining stars, then, before the whole universe turns black, the number of years that will have elapsed will be 10 to the power of 10 to the power of 76.

No, put away that pocket calculator. It cannot handle a number so vast. Indeed even to write the number down in the ordinary way, without using that convenient shorthand phrase 'to the power of', presents great difficulties. If we were to start with a 1 and then add all the necessary zeros and if each digit was no larger than a hydrogen atom, and we were to write down digits at the rate of 1,000 billion per second, then we would need no less than 200,000 billion billion billion billion Earth-sized planets on whose surface to write down the number.*

And the exercise could hardly be completed within a few weeks. Even with this accelerated writing, the task of writing Freeman Dyson's Number would take approximately 3,000 billion billion billion billion billion billion billion years, that is 3 followed by 66 zeros.

It is years that the number is intended to describe. But this is a mere convenience, for it really does not matter. What difference would it make if we were to pretend that it meant seconds, weeks, centuries or millennia? None at all. So majestic a figure

* *The billion here are 1,000 million.*

cannot be altered in any perceptible way by such paltry divisions and multiplications.

To return to the original question, assuming for a moment that we care about the lives of such extremely remote posterity, which, today, would we prefer to live in – a closed universe where it may be possible for intelligent life to exist for another 50,000 million years, or an open universe whose lifespan is governed by Freeman Dyson's Number?

The answer must surely be the second. Cassius made the point in *Julius Caesar:*

How many ages hence
Shall this our lofty scene be acted over
In states unborn and accents yet unknown!

On such a timescale, the glories and tragedies of humanity and its descendants would be virtually unending.

The messages from Saturn

IMAGINE a person standing on one of the moons of Saturn that have been discovered by the spacecraft Voyager 1: a rocky body some 50 miles in diameter, situated at the rim of the rings. He looks upwards towards the great planet, which fills half the sky – and he simply cannot believe what he is seeing.

A flat plain, 30,000 miles across, stretches vertically into the sky, apparently limitless on either side. It might seem that he could walk all the way to Saturn across this vast bridge in space.

This plain-like illusion, caused by the great number and density of rings, was perhaps the most astonishing discovery of the mission of Voyager 1 (and it is hard to write about its discoveries without using words like 'astonishing'). For there are no gaps between the rings – the so-called 'Cassini divisions'.

All the gaps are filled in with lesser rings, too faint to be visible from Earth. Seen from Voyager, the vast and beautiful ring system looks like a solid gramophone record, grooved with white and brown and grey, consisting of millions of rocky boulders and great chunks of ice.

Why this flattened, improbable shape? Why is Saturn surrounded by this great halo as if it were divine – as many ancient civilisations believed? Why have the rings not collapsed inwards, long ago, under the spell of the planet's gravity? And why are the three strands of one of the outer rings 'braided', or plaited like a girl's hair?

To these age-old mysteries there is only one likely solution. The newly-found moons, provisionally called S10 to S14, are holding the rings in position with their own gravity. In short, the 'rings of Saturn' do not belong directly to Saturn at all. Without the gravitational support of the tiny moons at their rims, the great system which has excited people's wonder from the Sumerian Empire onwards would scarcely be visible.

Even the strange nomenclature of Greek myth is not strange enough to explain the state of some of Saturn's larger moons. Most of them, indeed, Tethys, Dione, Mimas and Rhea, were what we have learned to expect from moons: cratered, barren objects, their surfaces smashed to rubble by the hammer-blows of meteorites through the ages. A few, however, were found to be inexplicably different. One side of Japetus is six times brighter than the other, for one side is covered with ice. Why? Where did the ice come from? Nobody knows. Surface ice suggests the presence of an atmosphere, and Japetus, with its 800-mile diameter, is far too small to have one.

Perhaps Arthur C. Clarke was wise to choose Japetus for his alien base in his film *2001: A Space Odyssey,* since conditions on this little world seem to defy rational explanation. So do they on giant Titan, the largest moon in the solar system, which is almost half the size of the Earth. Titan's dense atmosphere of nitrogen, with probable seas of liquid nitrogen and surface rainstorms of nitrogen droplets, is something so bizarre that no one, before last week, had given it serious credence.

Titan is a truly horrible world. With temperatures of minus 330°F, anyone or any object containing moisture that ventured

21

onto its surface – assuming that it has a solid surface – would be shrivelled to a skeleton within seconds.

To have found out all these things was a heroic exploit, but who was the real hero of the mission? The only possible candidate is Voyager 1, the spacecraft itself. It is such a sophisticated machine that it can make navigational decisions and even repair itself without reference to humans back on Earth, whose instructions would arrive too late, even travelling at 670 million mph, the speed of light.

Voyager's cameras and spectrometers (which identify material at a distance) are powered by a small nuclear reactor. The radioactive decay of a small quantity of plutonium provides the heat which keeps the rest of it operational.

In the long run, the construction of the two Voyager machines, with their small, on-board computers, may prove as important to man as the weird planets they were sent to investigate. A mere 12 feet in diameter, and weighing only 1,820lbs, these almost-intelligent robots will be the forerunners of smaller devices that will replace human divers in dangerous seas and miners thousands of feet below the ground.

The magnificent robot Voyager 1 is now lost to mankind for ever. Never again will it send us any signals. Yet its eon-long voyage will continue as it flies onwards towards the stars. A hundred million years from now, it will still be travelling away from us: battered, perhaps, by stray encounters with interstellar dust, but still recognisable. Imagine an astronaut riding upon it. He looks back to the solar system and sees – no sign of it. The great blazing Sun, nearly a million miles in diameter, is now too faint to be seen through Voyager's little telescope. Man, his home and his works, are all lost in the almost boundless immensities of the Milky Way.

Astronomy in the underworld

THERE is a dark world beneath a brilliant starry sky. The famous bright Sun is nowhere to be seen from it. The moon hangs motionless in the heavens, almost too dim to be seen but six times its normal size. And the planets? Look where you will, there is no sign of them.

Where is this gloomy region? Not on Earth but on the surface of the planet Pluto, the most distant known world in our solar system, 40 times further from the Sun than we are. Its discovery 50 years ago is now being celebrated by a new interest in far-off worlds and the publication of two excellent new books on Pluto and Charon, its newly-found moon.*

The detection in 1930 of far-off Pluto, a body no larger than our own moon and hence extremely difficult to find, makes a fascinating story. In 1929, Clyde W. Tombaugh, the son of a farmer who was too poor to send him to college, made himself a nine-inch telescope from old pieces of agricultural machinery and managed to get an assistant's job at Lowell Observatory, Arizona, where he embarked on his great search.

How would you or I set about looking for a planet more than 3,000 million miles from the Sun, beyond the orbit of Neptune, and whose very existence was doubted by professionals? At this distance, even if it existed, the Sun's reflection on it would be so faint that it would be indistinguishable from the faintest stars.

In what part of the sky should one search? And even if it were seen, how could it be recognised for what it was? After weighing these considerations, most people would abandon the quest in despair. But Tombaugh was very determined. He used a device called a 'blinker' which compared two photographs of the same small part of the sky which had been taken at different times.

Each picture might show anything between 50,000 and 500,000 dim stars. Now stars only change their positions (as seen from Earth) over periods of tens of millennia, and so any 'star' in those photographs which changed its position from one week to the next

---

* Out of the Darkness: the Planet Pluto, *Clyde W. Tombaugh and Patrick Moore (Lutterworth Press)*; Planet X and Pluto, *William Graves Hoyt (University of Arizona Press)*

could not be a star at all. It might be an asteroid, a comet, or perhaps even the long-sought planet.

It is a tall order indeed to examine the positions of hundreds of thousands of stars to see if a single one of them has changed its position from last night! But this is where the 'blinker' comes in. The two plates are superimposed on each other on a screen. If there is *any* difference between the two plates, the out-of-alignment 'star' will blink rapidly, drawing the astronomer's attention to it.

Even this piece of cleverness is not foolproof. In a dense starfield, there are many 'variable' stars whose brightness can change in a matter of days. Any such star, which seems to appear or disappear on the successive photographs, will be a suspected planet which, in detective parlance, will have to show a convincing alibi.

It took Tombaugh and his occasional helpers a full year before they found, in the constellation of Gemini, in an area that was only a 40,000th of the whole sky, a blinking object that had no alibi. It could not be a star, since no star had ever been charted in that position.

Several tests ruled out the possibility of it being an asteroid or a comet. It could only be the Sun's ninth planet, soon to be named Pluto after the god of the underworld, for it moved through regions of everlasting darkness, on the edge of the shoreless ocean of interstellar space.

An astonishing sequel to Tombaugh's discovery is that while it caused a world-wide sensation, Tombaugh did not at first get any credit at all. *'My colleague V.M. Slipher,'* he wrote afterwards, *'kindly cautioned me to beware of getting involved with out-siders who had only their own greedy purposes. There were greedy wolves out there. Packs of them. Right he was!'*

Bumptious professors, who had taken no part in the detective work, filled the media with their theories about Pluto. The phrase 'new planet' was on everyone's lips, but the name of Tombaugh was not. Still in his early twenties, he knew nothing of that great engine of publicity that should begin churning after a new discovery in science. As a result, his identity remained unknown to the general public for two decades.

Scientific reporting in those days, although considerably more verbose, was much worse than today's. On 14 March, 1930, a prominent London newspaper carried a lengthy account of Pluto's discovery. Its second paragraph was most revealing. *'The name of the astronomer has not yet been disclosed,'* said the reporter. *'But he is believed to be Professor Percival Lowell.'* A 'ghostly' discovery indeed! Lowell had been dead for 14 years.

Other solar planets may exist beyond Pluto, perhaps even giants like Saturn and Jupiter. But it will be very difficult to find them. The problem is not just the obscurity of distance, but the sparseness of the sunlight which they would reflect.

Yet there is hope. The Sun's effective gravitational field, which would provide the orbit for a Planet X, extends at least two light-years before it accedes to that of another star. Perhaps some modern astronomer, armed with the modern gadgetry which Tombaugh lacked, can discover that the gulfs between the stars are littered with barren worlds which our space-faring descendants will be able to use as fuel-dumps.

Waiting for the comet

Old men and comets have been reverenced for the same reason: their long beards, and pretences to foretell events.

Swift, *Works*

HALLEY'S Comet is on its way. In 1986 this most familiar to us of all those brilliant messengers from deep space, will be paying one of its once-every-76-years visits to the solar system.

The 1986 visitation of Halley's Comet may be less spectacular than its previous appearance in 1910. For most of the months it spends in our vicinity, it will only be visible through binoculars and telescopes, since it will be almost directly on the far side of the Sun.

But it will not, as in previous visits, pass by unstudied. Three satellites are going out from Earth to meet it: one, from Europe, made by British Aerospace and named Giotto, after the artist whose *Adoration of the Magi* showed a glimpse of the comet as it passed in 1301; another from Japan, and a third from the Soviet Union. America's NASA, to the sorrow and indignation of her space scientists, has been prevented by the Reagan Administration from playing any part.

This comet, first correctly identified by Edmund Halley in the seventeenth century, is but one of thousands of millions of 'dirty snowballs', lumps of ice, snow and dust which are part of the outer solar system, and of which a few thousand come close to the Earth at certain intervals on their long elliptical orbits.

Until Halley's famous and accurate prediction that the comet bearing his name would reappear in 1759, comets were considered wholly mysterious objects, having nothing to do with the orderly march of stars and planets, and probably bearing some ominous message to the rulers of the world. In the words from *Julius Caesar:*

> *When beggars die, there are no comets seen;*
> *The heavens themselves blaze forth the death of princes.*

Superstitions remain. Some fourteen visits of Halley's Comet have been witnessed during history – while thousands more must have gone unrecorded – and all of them have been associated, however foolishly, with calamities or great events. Here are some examples:

1057 BC: King Wu of China is engaged in a war, which presumably is calamitous for his enemies.

239 BC: Carthage, after being defeated in the First Punic War, is forced to cede Corsica and Sardinia to Rome.

78 BC: Marius massacres aristocrats in Rome.

2 BC: The Star of Bethlehem may have been Halley's Comet, but other theorists believe it was a conjunction of planets or a stellar supernova. In any case, Christ is believed to have been born in about 4 BC. The misdating of his birth was apparently an error made by Roman historians.

AD 610: The eastern Roman Empire is overrun by the Persians.

26

1066: William the Conqueror invades England.

1150: England devastated by civil war between Stephen and Matilda.

1378: Beginning of the Great Schism of the Papacy.

1531: Pizarro invades Peru.

1682: Turkey invades the Holy Roman Empire, starting a seventeen year war.

1759: Wolfe takes Quebec.

1910: Edward VII dies. Revolution in Portugal. Japan annexes Korea.

But these are meaningless coincidences, whatever people may have thought of them at the time, for Halley's Comet ignored many of the important dates in history. Appearing in 1456, it was a year late for the start of the Wars of the Roses and three years late for the fall of Constantinople. It was eleven years late for the signing of the Magna Carta, it missed the first Reform Bill by three years, and it was four years early for the outbreak of World War I. In short, the notion that comets predict or record human affairs on Earth is demonstrably absurd.

But things may be different in the future. Some astronomers estimate that, compared with the few that occasionally brighten our skies, there may be more than 10,000 million comets on the outer fringes of the solar system and bound gravitationally to the Sun.

Some of these are debris left over from the formation of Sun and planets some five billion years ago while others are from material gravitationally 'captured' during periodic passages of the solar system through the dusty spiral arms of our Milky Way Galaxy. According to the theory of the Dutch scientist Jan Oort, there exists a vast exterior belt of comets, Oort's Cloud as it is called, having a combined mass almost as great as that of the Earth, and extending out into space for a distance of one light-year, a quarter of that to the nearest star.

A good question to ask about any celestial object is: What use will it be? What use, therefore, is Oort's Cloud?

The answer is that it will one day assist our colonisation of space. For the nuclei of comets, their heads that shine so brightly when close to the Sun, are, on average, about 20 miles across. They are large enough to use as fuel dumps (since they contain complex chemicals) and as staging posts for far-ranging space-craft of the future. Many of them are even heavy enough to be used for a 'gravitational slingshot' – to accelerate a spaceship that swings around it and borrows some of its gravitational energy.

Since all nearby stars were formed in the same way and inhabit the same galaxy, it is likely that many of them are similarly surrounded by clouds of comets. If this is so, we can imagine a distant future when ships fly between the stars by 'comet-hopping'. Comets in the past may have borne us no message, but they may one day help bear us to the stars.

When the Earth is in danger

Who can be safe?
Guard as we may, every moment's a danger.

Horace

THE Earth is in continuous peril: not merely from human violence but from the violence of the universe itself. Gigantic explosions, outbursts of energy from the stars that could put a permanent end to life on this planet, could take place at any time.

A comet could strike the planet, killing tens of millions, a disaster only averted because it did not hit a populated area, when a large comet struck a remote region of Siberia in 1908. A stray asteroid, weighing untold millions of tons, might collide with the Earth, bringing consequent catastrophes of earthquakes, volcanic eruption, and the total blockage of sunlight through the raising of great quantities of dust.

The full rigour of the next ice age may ensue, according to the latest evidence perhaps within less than 20 years. How calamitous would be the effects on modern civilisation of the temperature

28

zones being turned within a few decades, into Arctic tundra! What of the fate of millions of refugees if northern cities were buried under hundreds of feet of ice?

There is nothing we can do to avert these disasters. They must have happened many times in the Earth's history, and they will happen many times in the future. While the chances of any of them occurring in the lifetime of anyone now living is probably small, thinking about the effects of possible cosmic catastrophes makes an interesting way to learn about science.

Let us take a look at supernovae, the total explosions which overtake some stars at the end of their lives.

The frequency of supernovae is about one per galaxy per century. Such an explosion is a spectacular object; for about three weeks before it fades, a disrupted star can shine forth with 100 million times its previous brilliance.

When such a catastrophe occurs more than about 50 light-years from the Earth (a light-year is the distance light travels in a year, moving at 670 million mph) it is nothing more than a spectacular curiosity. Such explosions were observed in our Milky Way Galaxy in 1054, 1572, and 1604. All of them occurred several thousand light-years away, and were important events in astronomical history.

But a supernova occurring less than about 50 light-years away would have lethal effects on us. Cosmic rays, charged sub-atomic particles released by the explosion, would eventually strip away the Earth's layer of atmospheric ozone, which protects us from dangerous ultra-violet radiation from the Sun. When exposed to it, an enormous number of people would die.

Now if a star is going to end its life in a supernova explosion when its nuclear fuel is exhausted, it must have a weight (or 'mass') about 3 times greater than the Sun's. There are 4 stars with a weight above this critical limit within 50 light-years of the Sun:

Star	Mass (Sun = 1)	Distance (light-years)
Procyon	3.1	11.4
Vega	3.1	26.4
Arcturus	4.0	35.9
Capella	3.1	45.6

Of even more terrifying proximity, with a weight 2.8 times greater than the Sun's, marginally below the critical level, is Sirius, only 8.7 light-years away. Fortunately, Sirius has a super-dense companion star known as Sirius B, whose strong gravitational field may in time draw matter away from giant Sirius and reduce its weight still further. If this happens, then the Earth will be safe from what would be the most destructive supernova of them all.

Maybe this natural safety process will occur. But then again, maybe it won't.

These deadly explosions may well not happen for hundreds of millions of years. Or again, they could happen tomorrow morning. The former eventuality is admittedly somewhat more likely than the latter, but predictions must be uncertain because so little is known about the precise age of these stars. Sooner or later the explosions will occur.

When a nearby supernova is observed, then mankind may have several decades in which to prepare for death. Cosmic rays travel at the speed of light, but never in straight lines. Instead, they move in erratic zig-zag fashion as they respond to every magnetic field which they encounter in space. And so, if we saw Capella exploding, we could count on anything between 10 and 100 years before we died in hundreds of millions.

The consequences of supernovae and other huge-scale natural disasters make interesting topics for discussion – with that hint of danger that makes them more exciting.

Hunt for a super species

WHILE argument goes back and forth about whether we are the sole inhabitants of our Milky Way Galaxy, whether there exist in it other beings, intelligent or not, nobody has suggested until now that to lack this information could be extremely dangerous.

In our present ignorance we are in the position of someone confined in a dark room, of whose geography and contents he knows little. The room may be full of treasures, all his for the taking; but it may also be the home of some predatory creature, with a vision as clear as his is opaque, which will devour him if he sleeps and ambush him if he ventures forth.

I take this idea, not from science fiction, but from the writings of a Nobel prize-winner in biology, Dr Francis Crick, whose book *Life Itself* contains one of the most extraordinary hypotheses about our origins that has yet been advanced.* His idea is this: that life did not arise naturally on Earth 3,600 million years ago. Instead, it came from space, and not by accident either. It was sent in the form of bacteria in a spaceship by members of an alien civilisation who, foreseeing their own doom, wished to plant their own seeds elsewhere.

It is a vast notion, and needless to say, there is no direct evidence with which to prove or disprove it. In the course of billions of years, all relics of the alien ship would have vanished. And, as Dr Crick admits freely, we do not know if the theory is even necessary as an explanation for the origins of terrestrial life.

For we do not know enough about the process in which our earliest single-celled ancestors are supposed to have formed from complex organic chemicals in the thin 'primaeval soup', the ponds that lay on the primordial earth. We do not know enough about the soup itself, or to what stimuli it was subjected, in the form of temperatures, gases and electrical disturbance by lightning.

In short, as Dr Crick states, *'our knowledge and imagination are too feeble to allow us to unravel exactly what might or might not have happened so long ago . . . . We cannot decide whether the origin of life on Earth was an extremely unlikely event or almost a certainty – or any possibility in between these two extremes'.*

Setting the scene in this modest fashion that avoids any dogmatic assertion, Dr Crick then speculates on a civilisation vastly superior to our own living more than 5,000 million years

* Life Itself: Its Origins and Nature, *Francis Crick (Macdonald)*

ago, that saw all its achievements threatened by some natural disaster.

Perhaps their sun, or one nearby, was about to explode into a supernova. Or perhaps, if they lived near to the centre of the galaxy, where the stars are more thickly clustered than out here at the periphery, a neighbouring star was on a collison course with their own. Or maybe they fell victim to a race-wide degenerate disease, as the development of their minds outstripped that of their bodies, and they forsaw that within a few centuries their whole cultures must perish.

Then, in the grand scenario envisaged by Dr Crick, they prepared colonies of bacteria which they propelled in the direction of thousands of likely stars, one of them being our own sun, then recently formed. All this they did, in the hope that on one planet at least, there would, over the course of eons, arise a species that was akin to themselves.

This fantastic proposition is certainly possible, but does it matter from any practical viewpoint whether it is true or false? We have colonised the Earth and wc may soon do the same to the solar system. Is it of any real importance to know how the process started?

Dr Crick's reason for indulging in such far-out thinking is to draw attention to the possibility of life elsewhere in the galaxy. It could, if it exists, be benevolent like his supposed super-civilisation; or it could be parasitical and predatory, a species resembling the horrible monster depicted in the recent film *Alien*.

His message is, that one way or the other *we ought to find out*. Searches of the sky, by all forms of astronomy of which we are capable, should be in continuous progress to solve this most fundamental of all questions. Until it is answered, man can know nothing for certain of his long-term future: whether he is free to do what he wishes in his own galaxy, or whether he is constantly at the mercy of a species more powerful and more deadly than his own.

But is anyone there?

THE nuclear physicist Enrico Fermi was famous for his three-word question which no astronomer could answer, either before his death in 1954 or now. 'Where are they?' he demanded, meaning those extra-terrestrial civilisations which, as all probability suggests, must exist in profusion throughout our Milky Way Galaxy.

Do they indeed exist in profusion, or do they perhaps not exist at all? For the most perplexing aspect of Fermi's question is the consideration that if they *did* exist, then some of them would have colonised or at least visited the Earth. But they plainly have not done so, and one must therefore conclude that they do not exist.

To avoid begging too many questions, I should explain that, in the opinion of scientists of many disciplines, it is inevitable that a super-advanced industrial civilisation, some centuries after attaining mechanisation, is bound to embark on a racial adventure of galactic conquest.

It is the opinion of these savants that sufficiently large spaceships can be built, with interior living spaces of many tens of square miles, so that inter-stellar voyages can last for many generations. In these conditions, it would take no more than about five million years to settle every habitable planet in the Milky Way.

Now five million years represents very little of the past lifespan of the Earth, and still less of our galaxy. Where, then, are these colonising aliens? What means this long delay? To be five million years late for an appointment suggests, at best, some carelessness, but a delay of 10,000 million years, the age of the galaxy, implies that no one is coming because no one is there.

The fascinating question is *why*. Why, apparently, is mankind the only intelligent civilisation in a galaxy of more than 100,000 million suns?

An original answer to this question has been recently suggested in that bold and speculative publication, the *Journal of the British Interplanetary Society*. In a word, man's uniqueness is because of the moon. But for the moon, it is argued, intelligence would never have evolved.

No, this argument has nothing to do with young lovers on summer evenings holding hands and gaping at the moon, and without a moon to gape at, there being no love and no procreation and hence the end of the human race. It is concerned, rather, with the vast tidal effects which the moon has had on the Earth.

The tides, although slight in their effect when we observe them from day to day, have tortured the face of the Earth, twisting its continents and oceans into shapes that would have been inconceivable had we lacked so huge a moon.

The Earth's surface may be said, roughly speaking, to comprise several land masses amid a vast expanse of ocean. But it did not begin in this way. It was slowly twisted into this unlikely shape by eon after eon of lunar tidal drag.

Without that tidal drag, calculations show, the Earth would either have been one solid land mass or else an unbroken ocean. In neither case would primaeval animals have been able to migrate from sea to land, thus hastening their evolution. Either of these conditions, therefore, would have been as unfavourable to the development of intelligence as the present circumstances favoured it.

It was a close run thing. The moon, once having been captured by the Earth's gravitational field, might have remained so distant that its tidal effects would have been negligible. If, on the other hand, its orbit had been much nearer to us than at present, that is to say within a critical limit of 9,600 miles, then it would have been broken into fragments by the Earth's tidal drag. We would be surrounded by a giant ring of boulders like Saturn. Sunlight would be sparse, our surface glacial, and of Nobel prize-winners there would be no sign.

Our chances of being so fortunate in the size and proximity of our moon must have been one in thousands of millions. Perhaps few, if any, other habitable planets in the galaxy have had such luck. Of our neighbouring worlds, Venus and Mercury have no moons, Mars has only a pair of tiny ones, and the moons of the giant planets, although large in comparison with ours, are far

34

smaller in proportion to their parent worlds. Locally, at least, the Theory of the Rarity of Large Moons holds true.

There are surely other civilisations in the universe: with a conservative estimate of one civilisation for every 10 galaxies, and an estimate of some 100,000 million galaxies, that would still leave us with 10,000 million alien civilisations. But they are probably too far away for any interesting contact to take place. The moon answers Fermi's paradox. Let us fear no space invaders! We ourselves will one day do the invading.

Beyond the dust in deep space

Darkness is more productive of sublime ideas than light.

Edmund Burke

FROM a 14,000ft mountain peak in Hawaii, British astronomers are seeing an entirely different universe from the familiar cosmos of brilliant stars. Instead of looking at very hot objects – which stars always are – they examine extremely cold ones, which have temperatures as low as minus 200° F.

This is the growing new science of infra-red astronomy, learning how the universe is constructed by peering, not at bright, spectacular objects, but at dark, cold things. It is a science at which Britain leads the world; for Edinburgh University's infrared telescope on the top of Mauna Kea mountain on the largest of the Hawaiian islands is the most advanced instrument of its kind.

Infra-red astronomy works in essence like this: every object that exists, even if it is only a few degrees above absolute zero (minus 459°F.), emits *some* radiation which may be detected by a sufficiently sensitive instrument. It radiates light in very long wavelengths, as opposed to those very short wavelengths which bright objects send forth. These long wavelengths are in the infra-red band of the spectrum, and an infra-red telescope detects them.

A good example may be found in the 'sword' that seems to hang from the 'belt' of Orion the hunter, where there is a huge gaseous nebula some 1,500 light-years away and several light-years across. To the naked eye it appears to be a single star, but through a good pair of binoculars it becomes a fuzzy, cloud-like object.

Embedded in this huge cloud are several 'proto-stars', stars in the actual process of formation. They are being warmed by the pressures of gravitational contraction, and several million years hence they will blaze forth in thermonuclear ignition to become true stars.

The temperature of some of these stars-that-are-yet-to-be is a frigid minus 10°F. Yet when seen through an infra-red telescope at a certain wavelength, they can be the brightest objects in the sky except for the Sun and the full moon. For many of these objects radiate thousands of times more energy at infra-red wavelengths than the Sun does at all its wavelengths.

How can they do so? At first sight it might seem impossible. But in this, they follow a nineteenth-century law, which scientists are only now able to exploit, that 'the total radiation emitted by an object is proportional to the fourth power of its radiation'.

Still more interesting things can be seen in the infra-red wavelength at the centre of our Milky Way Galaxy. This region, in the constellation of Sagittarius, is largely hidden from optical telescopes by immense clouds of interstellar dust. But infra-red telescopes can examine this cold, dark dust and show us what lies beyond.

And here, in the galactic centre, there are indications of a superdense object, an object so massive and with so strong a gravitational field that no light can escape from it – in others words, a black hole.

It has been suspected for several years that a black hole lies at the centre of our galaxy, devouring all stars that come near it. Its mass, fortunately, appears to be only about five million times greater than the Sun's, far too small to threaten the galaxy's stability, or to turn it into a violently implosive 'quasar'.

The astronomy of the infra-red will gradually reveal to us a new universe, rendering visible those things which our eyes can never see.

Dread of nakedness

Approach thou like the rugged Russian bear,
The arm'd rhinoceros, or the Hyrcan tiger;
Take any shape but that.

Macbeth

WHAT news of black holes? It has been some years since these objects burst upon public consciousness. Most people have now accepted the fantastic-seeming prediction of giant stars that have ceased to shine because their gravitational fields are so strong that no light can escape from them.

But black holes may yet have a surprise for us that is to be dreaded rather than hoped for. A situation could arise in which the discovery of one of these objects posed an actual threat to civilisation – not of a physical character, but rather from the information that it revealed.

Consider the two basic components of a black hole. Take first the 'singularity' within it, where the volume of the crushed star has been reduced to zero while its original mass has barely changed. It is a place of gigantic density in which not even sub-atomic particles can exist.

The second is the 'event horizon', that hides the singularity from outside eyes. Every object in space has a certain speed, an 'escape velocity', that must be attained if anything is to escape from it. From Earth, for example, a rocket must rise at 25,000 mph to defeat the pull of gravity and get into orbit. The greater a planet's mass, the higher its escape velocity. Now a black hole might have a mass three million times greater than the Earth's. Its escape velocity would exceed the speed of light, 670 million mph. Nothing, not even light itself, can normally escape from a black hole.

And so, as a black hole forms, an event horizon forms around it, and the light of its singularity, although still burning within, is to a distant observer snuffed out, like a car headlight extinguished in the night. All information about the singularity is hidden, only to be explored indirectly, by mathematical speculation.

But scientists see as a threat to their profession – yes, a threat – the possibility that in sufficiently extraordinary conditions, involving swirling deformities of mass, a singularity might form *without an event horizon*. The singularity would be 'naked'. We would be able to see into it. We could observe for ourselves the total breakdown of the laws of physics. We would see what Professor Paul Davies calls the 'edge of infinity'.*

Why would this situation threaten anyone? It is because *there would be no way whatever of predicting what would emerge from the singularity*. It would be the gateway into this universe from another, from a place, perhaps, where the laws of space, time and matter were different from our own.

The discovery of a naked singularity, says Professor Davies, would be a 'desperate crisis'.

'One can imagine', he explains, 'an Alice-in-Wonderland world in which the singularity coughs out all manner of weird and wonderful preformed objects, stars, planets, people, computers, copies of encyclopaedias. In a lawless universe (which this would be) anything goes.'

But how could this be? Surely, it will be objected, the singularity could only spew forth the relics of what had formed it. But this is not the case. The professor and his colleagues are explicit on the point. Up to the time when the singularity is formed, the behaviour of the doomed star is determined by prevailing physical conditions. But once the singularity takes shape, the influence of these conditions ceases. The thing becomes wholly alien.

* *Paul Davies, Professor of Theoretical Physics at Newcastle University, describes naked singularities and their consequences in his book.* The Edge of Infinity: Beyond the Black Hole *(Oxford University Press)*

If physical laws are seen to break down the reputation of physicists breaks with them. A world where events could not be predicted would be no place for rationality or logic. The hallowed principle of cause and effect would be destroyed. The scientist would have to step down from his lofty seat and surrender his place to the necromancer.

The last known naked singularity burst forth about 15,000 million years ago, and it did indeed emit marvellous objects, which later metamorphosed into typewriters, chessboards and princesses. I refer, of course to the Big Bang, the explosion from nowhere which began the universe. It was not only matter, space and time which started with the Big Bang. So also did order and logic; before it, none of these five things existed.

What happened once could happen again. Even on a small scale, a recurrence of the Big Bang could bring less welcome phenomena. There are many distant, violently-radiating objects in space that might turn out to be naked singularities. Let us hope none does.

Looking back to the Creation

There is occasions and causes why and wherefore in all things.
Henry V

AN astronomer was asked recently for his views on religion. He thought for a moment and replied: 'Ask me that question again in 1985 – or better still in 1986. With any luck, we'll all know a little more about the matter then.'

He was referring to the launching of the American Space Telescope, the first large optical telescope to be placed in orbit. At a height of 310 miles, free from the obscuring and distorting effects of the Earth's atmosphere, it will be able to see objects 50 times fainter than any yet detected. Its vision will encompass a

39

volume of space 380 times greater; and most remarkable of all, it will see seven times farther into the universe than has been possible from the ground.

Why, on hearing of such a project, should our thoughts turn to religion? Not least because an instrument of such power, able to peer through distances so vast, will be looking so much further back in time.

When we look at anything, through a telescope or through the naked eye, we do not see it as it is now, but rather as it was when its light started the journey to our eyes. No one on Earth has ever seen the Sun as it is, and no one here ever will. We only see the Sun as it was eight minutes ago, the time it takes for the Sun's light to reach us, travelling at 670 million mph, the speed of light.

Now the Space Telescope will see considerably further than the 93 million miles between here and the Sun. It will see no less than 14,000 million light-years, the distance light travels in 14

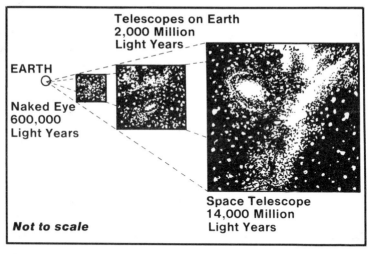

The naked human eye, on a clear night, can see about 600,000 light-years. The most powerful optical telescopes can see up to 2,000 million light-years. But the Space Telescope, to be launched by the space shuttle in 1985, will see seven times farther than this, up to 14 billion light-years.

40

billion years at the speed of light. We do not know how big the universe is, but we do know it is finite in size. Any mechanism which increases our vision sevenfold therefore brings us that much nearer to observing the Creation.

Here is the profoundest of mysteries. What 'caused' the Creation, the so-called Big Bang between 14 billion and 20 billion years ago, when matter, time and space came into being in one colossal explosion?

I put the verb 'caused' in quotation marks, since it would otherwise imply that the Big Bang was a consequence of something else. But this could not be; for in exploding, it created time. And so it is meaningless to say that anything could have happened 'before' it or have 'caused' it.

But this explanation is also unsatisfactory. It is a principle of physics that an event cannot happen without a cause. Yet here we have an event which *did* happen, but without any time being available for a preceding event to have caused it. The beginning of the universe, therefore, involves events that contradict the laws of physics as we know them, and that is why the Space Telescope invokes thoughts of the super-natural.

Three years ago, another astronomer, Professor Robert Jastrow, announced that he was embarking on a personal search for God by acquiring more information about the Big Bang.

Unfortunately, the forces of nature seemed to be conspiring to impede his investigation. The 'primordial atom' from which the explosion sprang was infinitely hot and infinitely dense. The explosion itself, if we are ever able to see it by looking backwards in time through the Space Telescope, will therefore be opaque, as if some form of cosmic censorship were at work. All clues to its nature will have melted in that holocaust. All, perhaps, save one. Those strange atomic particles called neutrinos, which penetrate anything, may form patterns that will tell us what really happened.

This is not to say that the Big Bang must have had a supernatural cause. It may be that different physical laws, amenable to investigation once we have the tools, take precedence in such extreme conditions over laws discovered and proved. But these things are still unknown. All we can do in our present ignorance is speculate that the Big Bang must have been an incursion from another universe, whose nature and very existence is beyond the realms of science.

The great cosmologist John A. Wheeler, has taken this speculation still further. It seems improbable, he says, that our universe emerged from a single 'other universe'. Why only two universes? Why not an infinite number? It makes more sense, in his view, to speak of a universe of universes, from which and into which all possible universes emerge and disappear. This place, if place it be, he called 'Superspace'. If universes were people, Superspace would be a womb for the new born and a tomb for the dead.

'Thinking about Superspace,' Professor Wheeler wrote once, 'is like chasing after Merlin. One moment it is a rabbit, and next a gazelle. And just as you reach out to touch it, it turns into a fox, or a brightly-coloured bird fluttering on your shoulder. It is the place where smoke comes out of the computer, because all classical laws of space-time break down.'

The mere dream of being able to explore these vast possibilities and many less exotic cosmic riddles – makes the Space Telescope the most exciting event in astronomy since Galileo.

SPACE TRAVEL

MANKIND is drawn to space colonisation, just as our ancestors felt compelled to explore the continents. If, as at present, the prospects for non-military space technology look bleak, then, as always, civilians will follow in the steps of the military. There are space stations to be built, moons to be mined and space colonies to be constructed.

Advancing with the military

WHAT effect will the Reagan Administration have on the growth of science? The question is important, since the whole world tends to decline or prosper in the long run on the productivity of American science.

In one sense, the Reagan Administration's science policy starts under the best of auspices, for it could barely be more catastrophic than its predecessors.

After a President who thought Venus was a flying saucer, who unjustifiably ordered the destruction of the space station Skylab, and who, while on a boating expedition, cried out that he was being attacked by a 'killer rabbit', we could hardly expect a deterioration.

Mr Reagan, it is true, does not seem to be a very scientific person. He has advocated the teaching of the Biblical creation myth in schools, and called evolution 'just another theory'. He and his wife regularly consult astrological horoscopes in the newspapers, and it took him many months to get round to appointing a Science Adviser in the White House.

But none of these seem very serious omens. Attacks on Darwin can be good vote-catching in the backwoods of America, and one can enjoy a horoscope without taking it seriously.

The President's first science budget, while it infuriated many scientists, should not necessarily prove bad for the growth of science. It resulted in the cancellation of many projects that would have produced knowledge and little else; but he accelerated others that, while aimed more at military prowess than knowledge, should, paradoxically, have the effect of vastly improving human welfare. Cancelled were several exciting unmanned missions to the edge of the solar system on the grounds that they could wait a few years, but he gave strong support to the manned space shuttle on the grounds of military security.

He cut funds for environmental protection, but he set aside millions for the production of nerve gas. He cut money for solar energy research by 60 per cent, but he authorised construction on America's only nuclear fast-breeder project, subjected, as it had been, to four years of financial harassment by President Carter. And he ordered the re-opening of a plant for processing nuclear fuel, which again, Mr Carter, in his nervous and baseless fears of nuclear weapon proliferation, had insisted on keeping closed.

In short, he stripped almost bare many of those public enterprises which might be of little help to the West in deterring aggression, but he let it be known that on defence expenditure there should be no limit.

Can civilian science and technology hope to flourish in such a climate? It most certainly can and will. Not, of course, to the benefit of bureaucrats and scientific administrators, who will lose their prestige or their jobs; but rather to the ultimate benefit of humanity.

Consider that most important historical precedent the Apollo rockets, which put twelve men on the surface of the moon. Did this begin as a civilian enterprise? Is it likely that it could ever have done so? Most assuredly not. The mighty Apollo launch vehicles were developed from Army rockets, whose original purpose was to kill people.

Some writers on space travel regard it as the greatest shame that their triumphs owe so much to military technology. For them it is a matter for lamentation that only the evils of the world make possible its most spectacular good works. How much more sublime and more beautiful, they say, would such achievements be if attained by single-minded altruism!

To me, these stirrings of tortured conscience are ridiculous. Dozens of people have flown in space; why should it so greatly matter how they got there?

So it will surely be with other systems of weaponry. The US Navy is taking old warships out of mothballs and equipping them with super-modern computer-controlled missiles. But, and here is the great hope, the computer programs that fire those missiles, if they are not to be useless, will have to be superior in cunning to anything which the Warsaw Pact can devise. What a chance for advances in computing! What marvellous new techniques in electronic software will eventually become available to those computer laboratories which are at present starved of government funds!

But it would take imagination, some people will complain, to see how the invention of newer and deadlier strains of nerve gas

45

will benefit anyone. On the contrary, it takes no imagination at all. It is obvious that the invention of a deadlier nerve gas requires a profounder knowledge of the nervous system than exists at present. And such knowledge, even if kept secret for a few years, may in the long run help to cure nervous diseases.

President Reagan has not 'destroyed' science, as some critics maintain. By shifting it from the civilian to the military sphere, he may even have given it greater impetus. This is a practice centuries old – since the Roman Republic built thousands of miles of military roads for the passage of its legions, which were then also used for the growth of prosperity through commercial traffic.

A feeling of space for the sightseer

MOST of the world's museums consist of objects surrounded by ropes with notices hung on them saying: 'Don't touch'. Very different are the space and rocketry museums at the Kennedy Space Centre in Florida and at Huntsville, Alabama.

These spectacular centres deserve to be visited by every British tourist going to the southern United States. If you have no chance of watching a rocket or a space shuttle lift off, the next best thing is to experience a simulated ride in a spacecraft.

This can be done with the most realistic effects at the Alabama Space and Rocket Centre, a few miles from Huntsville airport.

When the late Dr Wernher Von Braun and his crew of ex-Peenemunde technicians were at Huntsville designing the Saturn 5 rockets which took men to the moon, he was also at work on the construction of a museum that would both commemorate his fame for eternity and – a more amiable objective – bring home the feel of space travel to ordinary people through almost-direct experience.

Here, after walking through a hall hung with glowering portraits of the old German warlord, you enter a world of high technology marvels in which you participate directly, with no nonsense about not touching things.

'We wanted to give people the feel of what it was really like to fly to the moon,' said Museum Director Mr Edwards Buckbee, so he and his colleagues bought a cast-off 'Wall of Death' from a fairground, a large cylinder which spins so fast that people inside it are transfixed to its interior by centrifugal force.

They stick to its walls while the floor retracts because of the increased gravity created by the machine's rotation. Gravity and centrifugal force being the same thing, Mr Buckbee decided that a covered-over and darkened wall of death with couches and safety belts would portray very accurately what it felt like to be launched in a moon rocket.

It is most convincing. Gravity, two and a half times stronger than normal, presses on the body, exactly as it would in a real lift-off. And there are no special effects to give the impression that being hurled beyond the Earth's atmosphere to a speed of 17,000mph is far less frightening than television shots might suggest.

Simulating weightlessness is just as easy as producing additional gravity. It is done by swimming in fresh water with an aqua-lung.

Astronauts, later in this decade, will spend much of their time aloft, building large structures. Fixing together light metal beams so that they form the skeletons of space stations and riveting pieces of metal together in weightless conditions is surprisingly difficult. To train astronauts to do this while they are still on Earth, Huntsville's Marshall Spaceflight Centre has constructed a 70-foot tank filled with a million gallons of pure water containing all sorts of mechanical contraptions.

Dressed in full moon suits, as if they were outside their vehicles in earth orbit, astronauts descend into this tank and attempt, while upside down and at all sorts of angles, to perform specified tasks on different kinds of machinery. Weightlessness is exactly

47

reproduced by their wearing of the correct amount of weights while under water. It is fascinating to watch them do this through portholes outside the tank.

Even if one lacks the time to visit Huntsville, it is ridiculous to go to Florida and not make a trip to the launch pads of the Kennedy Space Centre, where there are bus tours twice daily.

Although the visitor will hardly notice it, there has been a great increase in security in this area since the days of the moon-landing missions ten years ago.

The reason is obvious. The Apollo Command modules had one purpose only – to go to the moon and come back – and, as a result, the Russians did not see them as a threat to their global ambitions.

But the space shuttle is a general purpose vehicle. It would be as useful in war as in peace. Soviet spy ships keep watch on everything that happens here.

The giant crawlers, for example, which carry the shuttle from the Vehicle Assembly Building to launch pad 39a at 1 mph, consuming 150 gallons of petrol per mile, are no longer kept together in a fleet, as they used to be. They are parked in widely separated places in case of a surprise bombing attack.

In 1971, I was allowed to drive my car to look at the rocket on the eve of a launch. Ten years later, when taking a look at the shuttle Columbia, I had to travel by bus in the presence of an armed guard.

None of this war-charged atmosphere need spoil the visitor's enjoyment. If anything, it makes things more exciting. This is a kind of sightseeing far removed from tours of traditional museums.

A manned trip to Mars

THE only American government official to my knowledge ever to advocate a manned expedition to Mars was the unpopular and later disgraced Vice-President Spiro Agnew. The effect was perhaps unfortunate. Since the day of his speech in 1969, at the time of the first manned moon landing, long-term plans to visit and colonise Mars have been left entirely to private research groups.

These Mars enthusiasts have long been active. It has been estimated that if all the different types of Mars spaceships proposed during the past 30 years were stretched end to end, the resulting contraption might reach half way to the planet of destination.

Now all this speculation has once more been made respectable by Mr James Oberg, a NASA mission controller and a distinguished writer on astronautics. In his book *Mission to Mars*,* he gives an expert's view of how and when such a mission should be carried out, and of the likely historical consequences of such a tremendous enterprise.

A vast project it will indeed be. Sending people to Mars and bringing them safely home again will be a far more complex task than arranging return trips to the moon. The sheer distance to Mars is the main problem. Consider the scale of it: at its nearest approach to earth, Mars comes no nearer than about 34 million miles, a distance 140 times greater than that between the earth and the moon.

At first glance, this might not appear to matter too much. With a maximum cruising speed of 25,000 mph, the Apollo astronauts took three days to reach the moon. At the same speed they could cover 34 million miles in just under two months.

But the situation is more complicated than this. People who set out directly for Mars, as if walking towards a distant mountain, will find that Mars is no longer there when they arrive. It will have moved away on its orbit. To catch up with this ever-fleeing world with the minimum use of energy, a Mars spaceship must itself fly in a vast orbit round the sun, aiming off, so to speak, with a voyage time of about seven months and a cruising speed of about 70,000 mph.

It might seem that a journey of seven months, with an ever-increasing lag in contact with Earth, might be a lonely business. There will be nothing to be seen from the windows except for the slow passage of some of the brighter planets, to indicate the ship's

* Mission to Mars *James Oberg. (Stackpole Books, Cameron and Kelker Streets, P.O. Box 1931, Harrisburg, Pa., 17105)*

speed. Some experts believe the voyage will be so monotonous, with possible outbreaks of neuroses or insanity, that the crew should be put to sleep for the duration of the flight.

Mr Oberg does not share this view. He believes – and as a mission controller he must know what he is talking about – that the crew would need to undergo constant training exercises that would prevent any possibility of boredom. The sheer complexity of the tasks they will have to perform on arrival on Mars will demand endless rehearsals.

And what is the purpose of this expedition, which will take two years and cost tens of billions of dollars? The first few visits, of course, will be exploratory and scientific. But in the long term, as Mr Oberg is careful to explain, *the plan is nothing less than the colonisation of Mars,* to turn this world of barren mountains and windswept deserts into an extension of Earth, providing a homeworld for millions of people.

This will mean 'terra-forming' the planet, to turn it into a place where people can walk and breathe without special suits and back packs. At present this is impossible on Mars, where average temperatures in Fahrenheit are 63 degrees colder than on earth, and the air density is a hundredth of ours.

But Mr Oberg, who has written widely on the theory of terra-forming, suggests that our great-grandchildren's generation could achieve this change by crashing asteroids into the Martian surface, gouging out huge craters, where, in localised areas at first, the air would be warmed by the energy of the collision and surface air pressure would increase in the low elevation of the craters.

'Plant and animal life,' he says, 'would spread across the new-barren surface, tingeing the red rocks with green and turning the red sky into a beautiful, beautiful dark blue. Liquid water would flow again on the surface, and the eons-dry channels and gulleys would become wet with new rains.' It is a splendid vision.

Moons as the mines of tomorrow

THE recent space adventure film *Outland*, set on Jupiter's moon Io, starred Sean Connery as an outer space sheriff in a celestial mining town, hunting down drug smugglers and being stalked by hired killers.

The script, unfortunately, was written before the two Voyager spacecraft flew through Jupiter's system of moons. For it is now clear that no sane person would choose Io as the site for any human activity! The character played by Sean Connery was very skilled at fighting gangsters, but even he might have been put off his stride on a world of continuously erupting volcanoes.

Io and her three sister moons, Europa, Ganymede and Callisto, together with mighty Jupiter herself, are the subject of a new book by Patrick Moore and Dr Garry Hunt, who is the only British scientist to have worked on the Voyager team.*

The existence of Jupiter's four large moons has, of course, been known since the time of Galileo, but only in the space age has their bizarre variety been revealed. 'Out there, says Dr Lawrence Soderblom, of the US Geological Survey, 'we've seen the oldest, the brightest, the darkest, the reddest and the most active bodies in the sun's family.'

Io, a roaring nightmare of a world, bears innumerable volcanoes, each erupting many times more violently than Vesuvius in all its fury. These unceasing explosions seem to be caused by Io's proximity to Jupiter, only 260,000 miles away.

Why this cause and effect? What does the proximity of Jupiter have to do with volcanoes on Io? The answer is gravitational. Suppose that our moon instead of being a mere 20,000 miles across, was the size of Jupiter with a diameter of 89,000 miles. How would this affect conditions on Earth?

They would become intolerable to civilisation. The tides, which now hardly trouble our lives, would transform the Earth's crust into a raging turmoil.

Let us seek a more agreeable world. Europa, a healthy 420,000 miles from Jupiter, shows none of these horrible tidal

Jupiter, Patrick Moore and Garry Hunt (Mitchell Beazley and the Royal Astronomical Society)

effects. Europa is so flat that it has been compared to a billiard ball. It has one very curious feature: it is criss-crossed with lines, so that from space parts of it look like a complex system of roads.

Further still from Jupiter, at a distance of 670,000 miles, swims huge Ganymede, the largest moon known to man, slightly bigger than Saturn's Titan. Now here is a mystery! Ganymede is covered with the crater marks of ancient meteoric bombardment, while Europa has almost none. Why this discrepancy? No one knows.

Callisto, 1.2 million miles from Jupiter, has been even more heavily battered with meteoric rubble than Ganymede. Its surface looks like the face of a person with an extremely severe case of smallpox.

But let us be practical rather than merely academic. What is the Jovian system *for*? How can mankind use it during the next few centuries?

One's first conclusion is that all the great moons, apart from Io, will be habitable. They are without oxygen and nitrogen atmospheres, but that can be remedied, by sealed domes with nuclear reactors to play the warming role of the distant Sun, and by underground habitation. Profitable mining and manufacturing operations will surely take place.

And Jupiter herself will not be neglected by tomorrow's industrialists. Its gaseous atmosphere, tens of thousands of miles thick, contains all the raw materials for hydrocarbon fuels. It is not too fanciful to predict that petrochemicals industries will eventually find their source of oil and gas in the atmospheres of Jupiter and Saturn.

Looking still further into the future, perhaps a thousand years or more, a still more advanced human race may begin to wonder whether the great bulk of Jupiter serves any useful purpose in its present orbit. Why not break it up into smaller fragments,. making hundreds of thousands of more manageable worlds, each pursuing its own path round the Sun?

This seemingly crazy idea was put forward 20 years ago by Professor J. Dyson, one of America's finest scientists. Later calculations showed that such a cosmic engineering project would be feasible, and the idea of dismantling a giant planet and putting its fragments round the sun is known as making a 'Dyson sphere'.

It is not merely to acquire useless knowledge that people send probes to distant worlds, however interesting that knowledge might be. It is exciting because we are shown something of man's future. Today's pure science is often tomorrow's industrial technology.

Space for Russian stealth

SOON after the maiden flight of the American space shuttle Columbia in 1981, the Russians temporarily suspended all manned space flights. Their reasoning was obvious. They regard the shuttle as a military vehicle (which it is only partly), and they need time to produce a craft of their own to match it.

Let us take stock of the Soviet space programme and see what the Russians intend by it, and how astronauts of the future can profit from its lessons.

It is a project that began as the personal propaganda toy of Nikita Khruschev. But since he was ousted in 1964 it has been almost entirely military in purpose.

Soviet military preparations in space, with experimental beam weapons with which to destroy Western ballistic missiles and perhaps even to attack targets on the ground, have been achieved more or less by stealth.

Khruschev's earlier taunts that Soviet successes in space provide Soviet superiority in almost everything goaded America to response. The landing on the moon, for example, might never have taken place but for the noisy campaigns that accompanied the flights of the Sputniks, of Gagarin and Titov, which were aimed at impressing the Third World and lowering Western morale.

Brezhnev and his colleagues were careful not to make this mistake. Accounts of space flights were kept factual and dry. No longer were the cosmonauts and their achievements eulogised.

Officials reported them in a routine fashion. Western journalists lost interest. Explorations of the unknown, which hitherto had half-covered our front pages, shrank to a few paragraphs among the rest of the foreign news.

Western public opinion assumed, wrongly, that the 'space race' was over. The error was made of thinking that what was dull could not matter. It was only by what seems now to have been the merest chance that the space shuttle project was started, challenging that dangerous Soviet domination of near-Earth space which might have led to Soviet domination of the Earth.

The Russians, now with nearly six years of uninterrupted manned missions behind them, have learnt many things about living in space which Westerners must re-learn for themselves because the Russians, typically, have been less than frank about their experiences.

As plans proceed towards the construction of orbiting 'battle stations', another problem perplexes the Soviet authorities – that of weightlessness on the human body.

Blood spreads more evenly through the body in weightlessness than it does when that same body is on the Earth's surface. The body 'thinks' that it has more blood than it needs, and produces fewer red blood cells and disease-fighting lymphocytes. As a result, astronauts become more vulnerable to infection.

The muscles, no longer needing to fight gravity, begin to weaken, and the bones lose calcium.

James Oberg, in his book *Red Star in Orbit,* quotes the in-flight personal diary of one Russian cosmonaut.

'Our faces have begun to swell', the cosmonaut writes. 'So much so that looking into the mirror I fail to recognise myself. I feel dizzy, nauseous. My movements lack co-ordination.'

I suspect there will prove to be one way, and only one, of countering these ill-effects, and that is to abolish weightlessness in spacecraft!

How? By making the spaceship rotate. As it rotates, just like a 'wall of death', in a fairground, gravity is created artifically, and the walls of the spaceship become its floor.

54

When gravity is created in this way, we call it 'centrifugal force'. But there is no good reason for such a distinction. As Einstein showed in 1916, gravity and centrifugal force appear to be the same thing because they *are* the same thing.

Here, then, are the two main lessons to be drawn. Weightlessness is bad for the health in many ways, and it must be abolished by having spacecraft that rotate. And their structure must be large enough, several hundred feet in diameter, so that in providing gravity they do not induce other uncomfortable effects.

And perhaps even more important for the short term, the engineering activities of a potentially hostile nation must not go unnoticed and without response because they seem dull.

To ignore the first lesson would be to abandon space flights for all save men of immense physical strength. And to ignore the second is to imperil freedom.

Room at the top

THE picture overleaf is of the giant space station that appeared in the 1969 film *2001: A Space Odyssey*. It is possible that the largest actual manned space station, that orbits the earth in the year 2001, could be as elaborate as this.

The space agency NASA, which constructed the American space shuttle, the world's most powerful and sophisticated spaceship that is now flying so successfully, desperately needs a permanent, and preferably manned, space station if the shuttle is to realise its full potential.

Shuttles without a station would ultimately be as useless as trains without platforms. The space shuttle by itself is no more than a vehicle for carrying people and cargo into orbit.

They need somewhere to go when they get there: a kind of central headquarters in the sky, an orbiting building which will combine the functions of warehouse, laboratory, factory, hospital, staging post and living quarters.

A very expensive project, one might suppose. Not so. The main functions of a space station are provided free by nature. It is kept permanently in place by the Earth's gravitational field, and its electric power comes from the Sun's radiation.

Building a space station will be relatively cheap in another, more subtle sense. Unlike the shuttle itself, we do not have to wait until it is finished before making use of it.

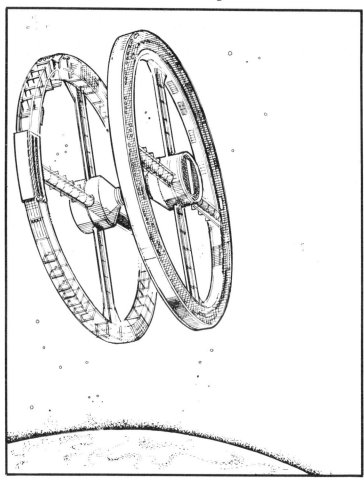

The giant space station from 2001: A Space Odyssey

It will be like a building that can be put to use as soon as the first foundation stone has been laid. The first task will be to erect a system of solar panels, about 300 miles above the Earth's surface, for electricity supply. Once this has been done, we will be able to say, in a sense, that the space station exists – even though this may sound like saying that a city has started to exist when it consists of nothing more than a socket for an electric plug.

The first step, then, is for a shuttle to take up some solar panels, packed up in its cargo bay, and assemble them in orbit. The next is to make use of the shuttle's main fuel tank, which is normally jettisoned immediately after launch (so that it falls wastefully back into the ocean). This is a towering structure, 154 feet high and 27 feet wide. When therefore, it has been emptied of fuel, and connected with the panels that collect solar radiation, it will be a fully electrified 'house' of some 90,000 cubic feet, the space of about 50 ordinary-sized rooms.

It will be a magnificent achievement at low cost; but even so, the modest space station that we have now, in imagination, created, bears little relation to the grandiose structure pictured left, with its twin wheels, each with a diameter of about 600 feet.

But the thing can grow in any direction and to any shape or proportion. It will continue its everlasting orbit unaffected by any lack of symmetry. And structures in space have yet another advantage over similar-sized structures on the ground. There is no weather out there, no wind, storm or earthquake. Building materials can be, so to speak, as light as gossamer. A structure that on Earth would be lucky if it stayed erect for an hour will survive in space for centuries as sturdily as the Tower of London.

At what point does a structure in space cease to be mere 'station', and start to resemble an orbiting city? It will not be long before the original structure, continually being added to as new functions are required of it, becomes immeasurably more complicated. Building in space will become one of the most fascinating tasks on which the human race has ever embarked. Two or more structures can be joined together. Other parts can be detached and sent to higher and more convenient orbits.

If we measure expense in energy consumed, the vastly greater part of it will be in bringing up materials from Earth. Assembling and reassembling those materials, once they are in space, will be accomplished at a fraction of the initial cost.

Yet there are some parts of the station that, in some degree, must always remain symmetrical. Look closely at the picture. The great wheels are obviously intended to spin. People living or working in their outer rims will be free of weightlessness. They will be subjected to centrifugal force, which is a form of gravity – and the gravity they experience will be proportional to the rate of the spin of the wheels.

By changing the speed of rotation of the structure, it will be possible, for the first time in millions of years, to vary the strength of gravity on the human body. What amount would people ideally prefer? Many doctors feel that the standard gravity found everywhere on the Earth's surface is too strong, that it is a main factor in limiting people's longevity by putting too much strain on their hearts. Some of the astronauts who walked on the moon, where the gravity is one sixth that on Earth, said afterwards that they had never felt better. By opting for a gravity that is somewhere between the Earth's and the moon's (which is weak enough to cause accidents), we may produce a healthier human species.

The ultimate step in the theory of space stations has been taken by a Princeton physicist, Professor Gerard O'Neill. He envisages, not mere cities, but colonies.

There may eventually come a time when it is recognised that living on the surface of a planet, where conditions are dictated by nature, is much less efficient and pleasing than building your own planet and choosing its conditions - like having clothes made to measure instead of buying them off the peg.

Such dreams must seem very distant to the Reagan Administration, now struggling with the question of when to start constructing a modest space station in a time of recession. Exactly when those first solar panels will spread forth above the Earth we cannot tell. But once they have done so, developments will be inexorable. It may prove as great a turning point in history as the emergence of our fish-like ancestors from the primaeval sea.

Two by two into the space ark

WHO shall inherit the heavens? The night sky has been searched intensively for signs of alien technology, and none have been found. No alien spaceships have been detected in our solar system, and there is no evidence that the Earth has ever been visited by intelligent extraterrestrial beings.

Nobody 'out there' has sent us a message or paid us a visit, and the reason, according to a growing number of scientists, is that nobody *is* out there. In our Milky Way Galaxy at least, we appear to be the only intelligent species.

Whether this belief is depressing or reassuring I will discuss in a moment. But consider the evidence for it and the implications for ourselves.

There is a tendency for life to spread into all the niches available to it, and therefore it seems axiomatic that a sufficiently intelligent civilisation will develop a means of interstellar travel – as our own species will certainly begin to do during the next three centuries.

Now until 1974, travel to the stars was thought to be overwhelmingly difficult. The reason was sheer distance. The nearest star beyond the Sun, Proxima Centauri, is 25 trillion miles away. At the distance, even if astronauts could somehow accelerate to a speed of 100 million mph (4,000 times the fastest speed of the Apollo moonships), a one-way journey would take nearly 30 years.

Such age-long journeys would be intolerably tedious, and people might well not care to undertake them. But this is only true if our ideas about spaceships are 'conventional' and based on the notion popularised in TV shows like *Star Trek,* that an interstellar journey must be completed within the lifetime of the astronauts.

Why not instead travel to the stars in a space 'ark', a flying colony, of the kind first proposed in 1974 by the Princeton physicist, Professor Gerard O'Neill. Such flying arks would be hollowed-out planets in their own right. They would be many miles long. Their inside surfaces, covered with an Earth-type environment of meadows and forest, would accommodate tens – or even hundreds – or thousands of people, who would live and die in them through many generations in the trackless paths of space.

If we can one day construct such space arks – and scientists are virtually unanimous in the view that we can – then intelligent aliens would have long ago done the same. The fact they have not done so appears to be strong evidence that they do not exist.

Yet the galaxy contains some 150,000 million suns. It may be thought a sufficiently large place for intelligent aliens to have started to colonise without yet crossing our path. But this reasoning is false. Consider the colonisation of the innumerable Pacific islands, most of which took place within a few centuries.

The rate of increase in the number of island settlements grew exponentially. From each island settled, new colonial expeditions set out to settle more islands, and from those islands in turn, fresh expeditions set out, and so on.

It can be similarly calculated that a species using space arks, and repeatedly constructing new ones from interstellar debris, could colonise the whole galaxy, from star to star, in the space of five million years. If this had happened, the Earth would itself have been colonised long ago. But this has not happened, and it is therefore a reasonable presumption that man is alone in the galaxy.

(One must not forget that there are thousands of millions of other galaxies. Whether intelligences reside in them, we may never know. For the scale of distances between galaxies is on average a million times greater than that between stars. Intergalactic travel, as opposed to interstellar travel, is perhaps too formidable an undertaking to contemplate. Besides, one galaxy is enough for one article.)

Significantly, many of the leading scientists who in the past fifteen years have written enthusiatic books about the prospects of communication with alien intelligence – Ronald Bracewell, Iosif Schlovskii, Benjamin Zuckerman, Patrick Palmer and Sebastian von Hoerner (although not yet Carl Sagan) – have now changed their minds. Absence of evidence may, after all, be evidence of absence.

If this is so, the Milky Way is ours for the taking. It is not yet teeming with intelligent life, but it will be. It will surely be found to contain many millions of uninhabited but habitable worlds. It is our destiny to go out and inhabit them, and no opposition need be feared.

One need only look through binoculars on a clear autumn night at the many coloured star clusters to appreciate their beauty like jewels in the sky. The simile is not inexact. Within a few million years, the same period of time since we evolved from apes, all the stars and all the great spaces between them could be ours. To paraphrase Marshal Blucher, what a place to loot!

ELECTRONICS

THE computer, a machine on which all high technology depends, once cost millions of pounds and was kept in a company's back room where it was managed by a mysterious priesthood called the Data Processing Department. Today, it is cheaply available to nearly everyone.

Silicon thieves

WHY should the Moscow Narodny Bank, acting in secret through middlemen, seek to buy an obscure little bank in California at nearly twice its market value?

An odd caprice indeed, it might seem. For the Californian bank was not a corporate concern: it did not control great industrial investments. It was just an ordinary drive-in-and-cash-your-cheque establishment.

The answer to the riddle is simple and ominous. This small American bank contained the personal and financial records of hundreds of engineers and scientists who work at nearby Silicon Valley, the largest centre of electronic chip manufacture in the Western world.

What opportunities for blackmail and theft this information would have provided for the Russians! How useful to know, for instance, that a particular scientist with access to top-secret laboratories was under paid and likely to have a grudge against his employers!

Large sums being paid to a local wine-shop? Why, perhaps the fellow is an alcoholic. Regular payments to some woman? Alimony, no doubt. Since divorced couples often hate each other, maybe the woman could give damaging information about her ex-husband with which he could be blackmailed. And here's a scientist who spends more than he earns. Perhaps, for a few thousand dollars, he would agree to rob his own company.

By such devious operations, the Soviet Union is seeking to build up its electronic weaponry. The attach on the Peninsula National Bank and two other Californian banks was foiled in 1975 when the affair was exposed by a Hongkong newspaper that is believed to have links with British Intelligence. Secrecy was blown and the plot was shattered.

But the Soviet policy of acquiring electronic components and know-how from the West for military purposes, both by theft and illicit deals, continues without cease. Computing power is needed to build radar, spy satellites, missile guidance systems and other offensive hardware. Yet rather than spend tens of billions of dollars to build up their own electronics industry, the Russians concentrate on purloining the goods of others, at a cost to them, since 1971, of only $100 million.

Let's take a look at some more of their methods. Soviet agents, like anyone else, can buy computers on the open market. But then they encounter a problem. For in most Western countries, Britain included, it is illegal to export electronic equipment without a licence, which is seldom granted for sales to the Communist bloc.

The way round this obstacle is easy. The Western dealer either conceals electronic packages in a suitcase or air freights large equipment disguised as air-conditioners, hi-fi goods or washing machines, and the false packages are shipped to Moscow.

Why do NATO countries have so many unscrupulous export dealers? The answer is simple: the fines, if you are caught, are tiny; and the profits, if you are not, are enormous.

I have not found a single case of a Western businessman going to prison for this virtually treasonable conduct. A typical case was of a Canada-based dealer who was fined $1,500 for selling electronic test instruments to the Russians which were worth $1.5 million. Who is going to refrain from crime when the penalty is 0.1 per cent of the sale?

Few people, in any case, are searched by the customs when leaving a country. (An exception was a raid on an Aeroflot jet at Dulles Airport, Washington, which yielded little more exciting – or more mysterious – than an American railway encyclopaedia.) In any case, the understaffed Customs officers of most countries tend to be more interested in the illegal export of arms, currency and drugs, than computer hardware which can be made difficult to recognise, or software which can come as a tape cassette, disguised as a musical recording, or even be written on paper.

The Western government agency which does the most damage through inaction is the US Office of Export Administration (part of the Department of Commerce), which has the gentlemanly habit of sending warnings to smugglers who are breaking the law, thus enabling them to escape. Two years ago, this accommodating agency announced that it saw no objection· to the sale of computers and automatic welders for the Soviet Kama River lorry plant, which provided lorries for the invasion of Afghanistan.

The unpleasant part of the story is that all this electronic equipment may be returning to the West, but not, unfortunately, in the way that it left. Using stolen or smuggled goods, the Russians have made their SS18 missiles into one of the deadliest ballistic weapons in the world, and have learned how to drop a warhead within a mile of an American Minuteman silo.

In the words of Mr John Lockie, a former US Attorney: 'We're going to see this stuff coming back at us. They're not making automatic popcorn poppers with it.'

Failing memories

Memory, of all the powers of the mind, is the most delicate and frail.

Ben Jonson

MANY people on reaching their forties suddenly find that their memories seem to be failing. They forget appointments, they cannot recall the names of people they recently met. It takes increasingly longer, on waking up in the morning, to recall things that happened the night before. Objects are more often mislaid. Places and street-names fade into oblivion. They conclude, falsely, that these lapses are the first signs of approaching senility.

But they are nothing of the kind. They represent only a natural weakening of that part of the mind, known as short-term memory, which was never more than a feeble instrument at best. It is true that all the information that has ever entered our brains is still there – somewhere. But that is small consolation if it cannot be recalled. Memory, therefore, is more usefully defined as the ability to recall information, and not merely the storing of it, a process that in most healthy people is automatic and unfailing.

This, indeed, is how science defines memory: 'A device for the storage and recall of information', whether the device in question is a brain, a filing system, a book, a library, or a computer. And one of the most difficult challenges of the twentieth century is to construct an artificial 'memory' that combines the best qualities of all these and eliminates the worst.

Such a device would be a computer program, but fed into a computer that would be far more advanced than present-day machines. The memory of a modern computer is immeasurably faster than a human's, but is so rigidly logical and mechanistic in its form that it can never be anything like so useful.

It wholly lacks our ability to associate apparently unrelated facts and create new information from them. Take the question: 'Who won the General Election of 1959?' A computer will either be able to answer this question instantly, or it will not be able to answer it at all. But a human, who cannot immediately remember the answer, may approach it in a roundabout way, something like this: The 1959 election? Let me see now. I was staying with the Joneses at the time, and I remember Jones getting frightfully angry because he had lost a bet on the result. And – of course! The winner was Harold Macmillan.'

To teach a computer to 'think' in this manner would be a magnificent achievement and scientists are working hard to this end. The main difficulty is one of hardware. The human brain, as might be expected from its millions of years of evolution, is superbly organised. Many of its tens of billions of neurons, or nerve-cells, are inter-connected. This enables a single incoming piece of interesting information to sweep through the whole brain.

But in the computer, by contrast, the separate chips are rigidly compartmentalised. Generally speaking, they cannot communicate with each other; and they are activated only when directly 'addressed' by the human user. The problem, therefore, is to try to redesign the memory functions of the machine to make them more akin to that of the human brain.

Humans, as far as we know, have three distinct memories: the immediate, the knowledge, for example, of whether a traffic light is red or green, a piece of essential but after all useless information which is almost immediately forgotten; the short-term memory – for a telephone number or an appointment tomorrow night; and, most reliable of the three, the long-term memory. This enables us to recall, always, and for ever, our names, our addresses, our nationality, and other data which are part of the bedrock of our personalities. A useful guide to these memories may be found in the lines from *Henry V*:

Old men forget; yet all shall be forgot,
But he'll remember, with advantages,
What feats he did that day . . .

All shall be forgot, for it is short-term memory, except 'what feats he did that day', i.e. how many Frenchmen he slew, how many he took prisoner and how much ransom he obtained from each. These are things so important, with so many 'advantages' for his own pride that they are recorded ineradicably in the long-term memory.

Weakening in middle age of the short-term memory should be regarded as having no more consequence than that weakening of the eyesight that often occurs at the same time in life. We may remedy the one by the purchase of spectacles and we shall, in time, with powerful electronic aids, find a substitute for the other.

Doomed publishers

IMAGINE a private library of the future that contains much of the world's literature, but only one book. Imagine its owner, who never troubles to visit a bookshop when he wishes to increase his collection, but merely signifies his wishes by typing out on his home computer terminal the number of the author and title, and his own credit card number.

Within a few hours, he has obtained a new novel, all 80,000 words of it, after it has travelled down a telephone line at the rate of several thousand words per second. Pleased with his purchase, he decides to take it on holiday, and so he tells his computer to store it all on a single 'capsule', a thing slightly larger than a button. This he takes with him, together with his 'book', a battery powered device about the size of a cigar box. Seating himself on the beach, he opens the 'book' and is confronted by a single 'page', in the form of a blank rectangular television screen.

He slots the capsule into the book, and the screen comes alive. On it appears the first page of a detective story with a line at the bottom that reads:

PRESS BUTTON A FOR NEXT PAGE

He goes on like this, pressing Button A whenever he has finished reading a page, so that the next appears instantly on the screen. To re-read a passage, he presses another button which instructs the machine to 'scroll back'. And by the same means, if he wants to skip, he makes the machine 'scroll forward', just as a present-day reader skips impatiently through boring passages.

The daylight fades, but he need not stop reading. He adjusts a knob and the type becomes larger and brighter. In short, whatever his natural pace of reading, whatever size of type he prefers, all his desires are accommodated. He has no need for the printed word.

Whatever can have happened to publishing companies that they distribute their wares in the form of electronically programmed 'buttons'? They will no longer exist. They will have become as redundant as those communities of monks in the Middle Ages who used to copy out manuscripts by hand.

Even today, the book publishing industry is doomed. Within a decade or so, authors will be able to dispense with its services. A few more years of progress in the miniaturisation of computers and the increase of information that can be put inside a single 'chip' will make modern publishing methods, such as the purchase of vast quantities of expensive paper, printing, binding, lorry transport and warehousing as archaic as the stagecoach and the bow and arrow.

It is an industry for whose passing few will mourn. Consider how incredibly time-consuming, expensive and inefficient is the process of publishing a book. From the time when an author makes the final alterations to his manuscript to the time when the book appears in the shops is approximately eight months. Compare this to a newspaper. The newspaper contains about three times as much information, sells at one-thirtieth of the price and is available to the public, not within eight months but within eight hours!

The executives of publishing companies often behave like the enemies of literature. They stand between an author and his readers. The sheer cost of publishing a book forces them to reject all but a fraction of the manuscripts they receive. There are thus huge numbers of authors, many of them people of great talent, who are unable to get their books published and are thus robbed of their livelihood.

It is perhaps salutary that electronic authorship should be on the horizon at the very time when the publishing executives have reached a trough of intellectual quality. This is an industry which almost deserves to be swept away by a technological revolution. Its capacity for judgement has fallen far since the nineteenth century when the work of an unknown author of richly complex fiction like Joseph Conrad was accepted by the first publisher who read it. Today, some people think that Conrad would be lucky to get into print.

Publishers will almost certainly be the first to take those fatal steps that will bring about their own destruction. Imagine the excitement in those boardrooms when it is learned that electronic writing, on tape or floppy disc, has at last become feasible. What a chance to eliminate those costly printers! What an opportunity to do away with those long months of waiting in which a book about current topics becomes out of date! And those lorry drivers. With computers and telephone modems we can transmit literature to the bookshops at the speed of light. Bookshops? What are we thinking of? We shan't need them either. We can sell directly with our customers by electronic mail order. We shall cut costs and production time by 90 per cent. Never fear, gentlemen. We'll soon make roaring profits!

While these hearty conversations are going on, authors themselves will be taking stock of the new situation. Their conclusions will be the same, but with one small difference. The publishers are not needed either. They, too, can go into limbo along with printers, warehousemen, lorry drivers and bookshop assistants. An author, from this time forward, will be able to write his book on his home computer (which will cost him about the same as a manual typewriter does today), get it reviewed if he can and advertise it on a computerised Prestel-type service to which all would-be book-buyers will have access.

Some marketing problems are still somewhat unclear, but authors will at last have the freedom of expression. They will publish whatever they choose and charge what they can get for it.

But not all authors are Conrads, and, quite obviously, this change will lead to an immense quantity of rubbish appearing on the market. What of it? It does not matter. If a book is no good, or if an author has a reputation for writing bad books, then he won't make a living. The rules of the book market, like any free market, will drive out the bad and encourage the good.

It will be argued that the quality of writing will suffer without the editing service which publishers today provide. But this need not happen. There is one group of people in publishing firms who will survive. These are the copy-editors, who can turn an indifferent manuscript into a good book by suggesting thousands of relatively tiny improvements that make up one vast change for the better. A new profession will therefore arise: that of private 'consulting editors', people like doctors and dentists, to whom authors will take their manuscripts for editing before publication, in exchange for a fee.

Do not weep at the thought of those publishing executives losing their occupations – for they do so in the fine cause of a freer literature. Look at it this way: if they are people of talent, they will soon find re-employment. But if they are not then perhaps they should never have been employed in the first place.

SCIENCE AND WAR

THE horrors of the battlefield will probably last as long as the human race. The Soviets are reviving ancient chemical poisons on Earth, while our moon will sooner or later provide the site for the ultimate missile base of one or more of the Superpowers.

Yellow rain

He will practise against thee by poison.

As You Like It

FOR centuries, communities from Europe to Asia died in agony in huge numbers when their bread became polluted by virulent fungus poisons. Soviet scientists isolated these poisons in the 1930s, and have since been mass-producing them as a means of mass murder.

Most of the advanced nations, it is true, either manufacture or carry out research into chemical weaponry. But the Soviet Union and its allies have outstripped all others in the intensity of their devotion to the development and use of poison.

What are these substances? The most lethal toxins used in modern warfare are still the hideous natural poisons that one associates with the Dark Ages, rather than any synthetic material created in the laboratory.

Democratic countries have been pitifully slow to recognise and counteract the advances which Eastern dictatorships have made in this field. It comes as a dark surprise to today's Western mind that the technological societies of the Communist bloc are but a veneer on a base of mediaeval barbarism, in which poisons extracted from herbs, fungi, snakes, amphibians and fishes are often the most favoured way of getting rid of an enemy.

It was in this tradition that the Soviet Union began its 1980 invasion of Afghanistan with the most terrible arsenal of offensive chemical weapons used by any army in history. Countless Moslem rebels died in convulsions from attacks by clouds of 'yellow rain'.

Nor should there be too much surprise at the manner of their death. To quote from an excellent book on chemical warfare, 'the Red Army demonstrates a military psychology that makes it possible to use war poisons without hesitation, as simply another weapon.'*

Let us look at the history of one such poison: ergot, a fungus toxin which has been known for nearly 3,000 years. An Assyrian

* Yellow Rain: A Journey through the terror of Chemical Warfare, *Sterling Seagrave (M. Evans and Co., New York)*

74

tablet of 600 BC first mentions it as a noxious pustule found on ears of grain. It probably caused the plague which nearly destroyed Athens during the Peloponnesian Wars, when starving people were forced to eat bad bread. It caused mayhem in Duisberg, Germany, in 857 AD, and in wide areas of France in 943.

A French chronicler of that year speaks of people 'shrieking and writhing, rolling like wheels, foaming in epileptic convulsions, their limbs turning black and bursting open.' Then he explains: 'The bread of the people of Limoges became transformed upon their tables. When it was cut it proved to be wet, and the inside poured out as a black, sticky substance'.

The cause of these horrors which became endemic among the ignorant peasantry was bad harvesting and grain storage, that permitted fungal growths on bread. Ergot, and similar fungal poisons, specially treated in Soviet laboratories, are nowadays used against rebel villages in Laos and Afghanistan, as Mr Seagrave's book reveals in detail.

For mass killings or for individual murder, the ancient poisons are proving most efficacious. The Bulgarian exile Georgi Markov, hated in Sofia for his BBC broadcasts, was murdered in London in 1978 by an agent using an umbrella tipped with ricin, from the castor bean, which the murderer had boasted in a telephone threat to Markov 'is a poison the West cannot detect or treat'.

The greatest danger of all is that some group of ill-intentioned people might seek to combine the ancient poisons with the techniques of modern science to create a new weapon of unprecedented frightfulness.

It could happen like this. Genetic engineering, the laboratory manufacture of microbes through the alteration of genes, promises much for better medicines. But this hopeful new technology could be perverted to make a 'monster microbe' that would colonise the human intestine with 'pili', or tentacles, with which to adhere to its walls. For such a poison, there might be neither treatment nor antidote, and anti-bodies would accept it as being normal. A vial of it dropped in the water supply of a few major cities could, within days, produce a catastrophe to rival the Black Death.

One scientist who has warned of just such a danger is Professor

Donald B. Louria, of the New Jersey Medical School. Explaining his worst fears, Professor Louria has said: 'One microbiologist with whom I discussed this scenario said it could not happen because the experimenters themselves could not avoid becoming victims.

'But this is nonsense. They could immunise themselves against pili before the toxins were added, so that the bacteria could not take hold in their intestinal tracts. I believe there are those among us on this planet so venal, so committed to achieving power, or simply so mentally warped, that they would do exactly as I have outlined.'

One doesn't have to be a geographical genius to predict just who these people might be. That is, if they thought they could get away with it.

Space Wars

THIS decade is likely to present greater dangers to mankind than any since the end of World War II. If the Soviet Union succeeds in placing an operational laser battle station in orbit while the Americans fail to do the same, the free world will be at the mercy of its enemies, most of its strategic weapons rendered useless.

The reason is simple. A laser beam fired in the vacuum of space can, or will soon be able to, punch fist-sized holes in metal objects at a range of hundreds of miles. This means that American intercontinental ballistic missiles, which make some of their journey through space, could all be destroyed before they reach their targets.

Nor will Western missiles that travel to their targets without leaving the Earth's atmosphere, like the Cruise and the Lance, be necessarily safe from enemy battle stations in orbit. While the energy of laser beams can dissipate in air, especially on cloudy days, this is not true of weapons which shoot beams of charged particles.

Polaris submarines will soon be at risk from spy satellites. For many years they have been safe in the secret depths of the oceans, able to inflict more damage on the Soviet Union in the space of four minutes than Hitler did in four years. But this is unlikely to be true for much longer. The Russians have a large and growing fleet of space-borne anti-submarine satellites, with a developing ability to detect the infra-red 'scar' which a submarine leaves on the surface, enabling them to track its movements.

In short, with space warfare, strategic weapons are entering a new realm of technology. Thanks to four inactive years during the Carter Administration, the Russians have gained a substantial advantage in their efforts to acquire the ability to destroy Western strategic forces totally and without warning. Unless America acts with determination, we may be faced in this decade with the choice between surrender or destruction.

Not being privy to the councils of the Pentagon, we cannot be sure whether the Americans are reacting to this crisis with sufficient speed and vigour. It is only possible to be certain of one thing: that the space shuttle, a quarter of whose flights will be military in purpose, will add enormously to America's ability to place weapons in orbit. And weapons there are needed above all else.

Only if the new Soviet threat is successfully countered can there be hope of continuing the mutual balance of terror, which has prevented war between the super-powers for more than 30 years, and which now is trembling so dangerously.

The old balance, consisting of thousands of missiles in their silos, will give way to dependence on electromagnetic weapons which move their targets, not at a cumbrous 17,000 mph, but at the speed of light, 670 million mph. This, like previous great advances in military technology, is likely to lead in turn to new social developments. Let us try to predict what they will be.

The first consideration is that the existence of opposing laser battle stations in orbit, each holding the strategic forces of their client state in pawn, will not be the end of the cold war in space. Battle stations can themselves be attacked, and those weapons which threaten them will in turn be vulnerable to assault. The race will be on to construct the 'ultimate' space weapon, a battle

77

station so powerful and with such impregnable defences that all objects in low Earth orbit will be at its mercy.

One of the two safe places to instal such a weapon will be beneath the surface of the moon. On the moon? At first sight, the idea must seem crazy, but it is being seriously considered as a long-term contingency plan by specialist groups at the Redstone Arsenal in Huntsville, Alabama, and at Strategic Air Command in Omaha, Nebraska.

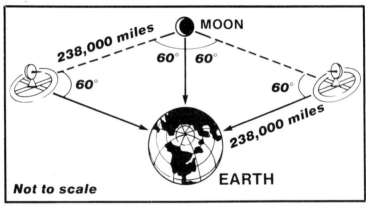

Laser guns on the moon and in Lagrangian orbits

Consider the advantages of a manned lunar laser battle station. The only remaining technical obstacle is the creation of a laser with sufficient power and narrowness of beam to destroy space vehicles at a range of 238,000 miles. But once installed it would be almost impossible to find, since it could be hidden anywhere among the moon's craters and canyons. It could not be destroyed by an opposing laser, since the enemy would not know where to fire. Nor could it be immobilised by a nuclear missile, since the approaching warhead would itself be vulnerable to the laser.

Building the station will, of course, require considerable preparations which can be observed by telescope. Would this reveal its intended location? Perhaps not. We speak now of a period 20 to 50 years hence, when civilian activity is likely to be taking place on the moon on a large scale. In this situation, military construction can be concealed. Peaceful technology is

78

likely to follow the military lead into space, as it has in so many fields. As in the empires of old, the merchant will walk in the tracks of the army.

But the lunar battle station will have one disadvantage. It will only be effective in deterring aggression for about half the day. Anyone can verify, by playing with a small globe, that there are several missile flight-paths between Russia and key Western targets which, for some parts of the day, will not be in line of sight of the moon.

A Superpower desiring absolute command over the Earth would therefore need at least two more battle stations in deep space, so that all parts of low Earth orbit could be covered round the clock.

Where should they be placed? The ideal locations would be in two out of the four Lagrangian orbits.

The French Comte de Lagrange made in 1788 one of the few remarkable mathematical discoveries about the universe that took place between Newton and Einstein. Around two orbiting celestial bodies – in this case the Earth and moon – there are four points at which a third body could form an equilateral triangle with the other two, and remain there forever in stable orbit. The behaviour of objects in these locations, would be influenced equally by the gravity of two worlds, providing stable vantage points for battle stations.

The pattern of war, and of preparations for war, may be extended ever more deeply into space in the distant future – with all man's activities – until the Earth itself ceases to be the target and the prize.

But whether we can survive to inhabit that distant future depends on decisions being made now; on recognising that Earth-bound weapons will soon no longer deter aggression, and on deciding swiftly what to do about this fact.

MATHEMATICAL ENIGMAS

C LOSELY related to computer science are the marvels of mathematics. Here we have some explanation, if an explanation is possible, of the mysteries of infinity, of magic squares and of the remarkable meditations of Professor Douglas Hofstadter. Douglas Hofstadter.

For ever and ever

How can finite grasp the infinite?

Dryden

FEW people have ever learned mathematics, even at the most elementary level, without wondering about infinity: a number so big that it is bigger than any number we can think of.

The very thought of infinity can provoke nightmares, dreams of experiences that are horrible because they are endless and therefore meaningless: a loop of thought that continuously returns to itself, a corridor which has no end, a library infinitely full of books. Even the perpetual crash of waves on the seashore has in it a sort of horror – because it will last as long as the world and still carry no message.

Infinity is all the more tantalising because it does not seem to have anything to do with the real world. Some very large numbers emerge from natural science, but they are all finite. The stars in the heavens may seem uncountable; but if current theories of cosmology are correct, they total 'only' about 10 to the 22nd power (1 followed by 22 noughts). Even the estimated number of atoms in the universe, which might on first consideration seem unimaginably vast, is in fact a quite manageable number, which any decent pocket calculator can handle, 10 to the 80th.

The number of possible arrangements of these atoms, the factorial of 10 to the 80th, that is to say, the product of all numbers between this number and 1, is of couse not an easy quantity to wrestle with, but let us pass that by. The next largest interesting figure is the haemoglobin number.

In haemoglobin, which is the chief protein component of red blood cells, there are 539 different amino acids. The possible number of rearrangements of these amino acids, a matter of some interest to biochemists and biologists, is therefore the factorial of 539, roughly speaking 4 followed by 619 noughts. A big number indeed, but are we to stop here? Someone interested in numbers for their own sake might want to know *the factorial of the haemoglobin number,* the index, so to speak, of possible rearrangements of rearrangements. This would equal . . . well, on second thoughts let's pass that one too.

But we are far short of infinity. Obviously, there is an infinity of whole numbers. The series, 1, 2, 3, 4 . . . can go on for ever. But there are many other 'sub-sets' of infinities. And perhaps, as some theorists maintain, an infinity of infinities. Prime numbers, for example, numbers only divisible by themselves and 1, like 2, 3, 5, 7, and 11, are also infinite in number.

So are the number of digits to the right of the decimal point in pi, the ratio of the circumference and the radius of a circle. The first million of them were published in 1973 by two French mathematicians in what has been called 'the world's most boring book.' The book went like this: pi, it said, equals 3.1415926538 9793 . . . and, omitting the next 400 pages and 999,975 places . . 5779458151. The General Assembly of Indiana had decided in 1897 to cut out all such nonsense by enacting that pi was equal to 4. They should have been made to travel in carriages with elliptical wheels as a punishment.

A startling discovery about infinity, which many people at the time refused to believe, was announced in 1874 by the German mathematician Georg Cantor. His Continuum Theorem states that infinities come in different sizes.

What he proved was this: that the quantity of 'real numbers' was greater than the quantity of all 'natural numbers'. (A real number to a mathematician, is any number, whether it has fractions or not, like pi, or 2.4, while a natural number is a whole number, like 1 or 2.)

Common sense tells us that this is likely to be the case, since, after all, between the natural numbers 1 and 2 there is an infinite number of real numbers. But proving it must have been a very different matter. How can one prove that one sort of infinity is greater than another sort, when both sorts are infinitely large? Anyway Cantor did prove it, but I make no claim to understand his proof and those seeking further elucidation should consult an advanced text book.

Here is an interesting problem that readers might like to solve:

If there were an infinite number of worlds, why would there not be an infinite number of Mrs Thatchers?

This is your postcard speaking

THE American mathematician Douglas R. Hofstadter was intrigued recently to receive a postcard from Britain bearing the Post Office imprint: Be properly addressed. 'Was this an order issued to the postcard itself?' he asks. 'If so, then British postcards must be more intelligent than American ones. This is the first time I have met a postcard that could read, let alone correct its own address.'

One's first reaction is to smile at this nice sarcasm against our Post Office and think no more of it. But Professor Hofstadter is far from being a sarcastic humorist. His speciality is 'artificial intelligence', which means trying to build computers which can think, and he is very serious about the matter.

One of the obstacles to this enterprise, in which governments around the world have invested hundreds of millions of pounds, is that it has so far proved extremely difficult to teach computers to understand the meanings of ordinary English sentences.

The rules of grammar and punctuation they can learn, since these are more or less explicit. They can correct misspellings. They can replace cumbrous phrases with simpler ones. And they can even, by a combination of searching for 'keywords' and parsing the sentences which contain them, decide whether a piece of prose is 'interesting'. But of its deeper meaning they have no notion – any more than they have any awareness of their own existence.

Professor Hofstadter, in the January 1982 *Scientific American,* muses on this matter by playing with sentences which seem to be aware of their own existence. For a sentence has this in common with a computer program: it is merely an arrangement of symbols which conveys a deeper meaning.

But how can a sentence 'be aware of itself'? And is Professor Hofstadter being serious? It is hard to know, and perhaps it does not matter. After all, great problems of science have often in the past been solved by fooling around with words. Anyway here is an example he gives of a sentence which is entirely self-pre-occupied:

In this sentence the word and *occurs twice, the word* eight *occurs twice, the word* four *occurs twice, the word* fourteen *occurs four times, the word* in *occurs twice, the word* occurs *occurs fourteen times, the word* sentence *occurs twice, the word* seven *occurs twice, the word* the *occurs fourteen times, the word* this *occurs twice, the word* times *occurs seven times, the word* twice *occurs eight times and the word* word *occurs fourteen times.*

Or alternatively:

If a dog had written this article, he might have said that people are inferior because they don't wag their tails. This gave me paws for thought. What might this article have been like if it had been written by a dog? I cannot say for sure, but I have a hunch it would have been about chasing squirrels. And it might have had a paragraph speculating about what this article would have been like had it been written by a squirrel.

More subtle, but just as self-preoccupied, is the eighteenth century Dutch story about Liars' Bridge. A variation on the famous statement of Epimenides the Cretan that 'all Cretans are liars', it goes like this:

A man is out walking his son, and the son tells a whopping lie. The father rebukes him sternly, saying that they are approaching Liars' Bridge, which always collapses when a liar tries to cross it. The boy confesses he lied, and they cross the bridge. But the bridge collapses because the father lied about it being a liars' bridge.

This story, in turn, is reminiscent of Bertrand Russell's paradox about a barber in a certain town who announces that *'he shaves all men who do not shave themselves'*. Who then, shaves the barber?

Here are more self-referential sentences collected by Professor Hofstadter, many of them also contradictory or paradoxical:

'The reader of this sentence only exists while reading it.'

'When you are not looking at it, this sentence is in Spanish.'

'If this sentence was in Spanish, it would say something else.'

'This sentence will end before you can say Jack Rob . . .'

'If I had finished this sentence . . .'

'Does this sentence remind you of Agatha Christie?' (How

could it? It doesn't say anything about her. But at the same time, in a curious way, it does remind us of Agatha Christie.)

The professor concludes with a discussion of the famous cigarette advertising slogan:

> *'Winston taste good like a cigarette should.'* This surely has an unintended meaning. For the observation that a cigarette *'should taste good'* implies that many brands do. And if that is so, what makes Winston so special?

Most scientists are far too reserved to put their near-frivolous meditations into print in the way that Professor Hofstadter does. It is perhaps their loss – and ours.

Magic squares — and other marvels

Mathematics may be defined as the subject in which we never know what we are talking about, or whether what we are saying is true.

Bertrand Russell

LIKE the 'intelligent postcard' of the last chapter, this article has revised its own first two paragraphs. Originally, it was to draw attention to a 'Roadshow' being held by the London Mathematical Society, at which distinguished mathematicians would lecture on the more obscure aspects of their craft.

Unfortunately, the article below sorely embarrassed the Society's officials, since it discussed mathematics at a much lower level than that expressed by the lecturers. Bus loads of number-enthusiasts arrived, causing an overflow and jamming the lecture hall. They went away highly disappointed at not understanding a word of what they had heard. Anyway, here is the article which made the chiefs of this learned Society somewhat regret having asked me to publicise their proceedings in advance.

W HAT do people do in this strange world of fundamental or non-applicable mathematics?

This kind of mathematics is seen by many as a ridiculous occupation, and those who indulge in it are sometimes regarded as amiable lunatics. Long remembered will be the *New Yorker* cartoon of a few years ago which showed half a dozen grey-bearded men staring in puzzlement at a huge blackboard covered from side to side with abstruse equations. At length one of them remarks: 'Surely, aren't eight nines seventy-two?'

And a profession which depends so much on the rigours of logic can be an easy prey for satire. Consider this anonymous proof that every horse has an infinite number of legs.

Horses have an even number of legs (so runs the proof). *Behind they have two legs and in front they have fore legs. This makes six legs, which is certainly an odd number of legs for a horse. But the only number that is both odd and even is infinity. Therefore horses have an infinite number of legs.*

More seriously, the exploration of numbers is a fascinating way to pass the time. Let us take a look at some peculiar numbers.

The number 37, for example has some very odd properties. One would not expect that the sums 3 x 37, 6 x 37, 9 x 37, 12 x 37, 15 x 37, 18 x 37, 21 x 37, 24 x 37, and 27 x 37, would yield any especially surprising answers. But they do. The products are, respectively: 111, 222, 333, 444, 555, 666, 777, 888, 999.

Palindromes, numbers like 2772 that read backwards exactly as they read forward, have long intrigued mathematicians. Here is a method of generating numerical palindromes which, for reasons which nobody understands, works in most cases but not always:

Take any number with two or more digits, let us say 56. Reverse it, making 65. The sum of the two numbers is the palindrome 121. The number 56 is remarkable in producing a palindrome in only one step of addition. The number 319 takes two steps to produce the palindrome 3553; 571 takes four steps to become the palindrome 9559.

There are many mysteries about this procedure. It has never been proved that it will produce a palindrome from any number of two or more digits – nor, for that matter, has it ever been disproved. Some numbers take an enormous number of additions, some take only one or two, and the number 196 does not produce a palindrome no matter how long one keeps trying; or at least, no one has succeeded in making it do so.

Many games require the use of random numbers, numbers which bear no relation to each other and are produced, so it seems, purely by chance. A pocket calculator provides an easy means of generating random numbers. The trick involves constant multiplications by 147.

It works like this. Write in the display of the calculator any number between 0 and 1, using as many decimal places as it will take: for example, 0.9535731793. Multiply this number by 147, giving 140.1752574. Subtract the whole number 140 from it, leaving 0.1752574.

Drop the decimal point, and keep only the first four digits 1752. This is the first random number. Now repeat 0.1752574 by 147, and subtract the whole number part (which in this case is 25). The second random number produced is 7628. Continue like this, assuming your calculator has a display that will accept numbers of six or seven digits, and you will have the following table of random numbers:

1752, 7628, 1362, 0257, 7848, 3747, 0917, 4816, 8004, 6608, 1394, 5056, 3301, 5279, 6153.

In Lennart Rade and Burt A. Kaufman's *Adventures with your Pocket Calculator,** a book wholeheartedly to be recommended for people wishing to use these little machines for purposes more elevated than adding up expenses, there are many other absorbing exercises.

The study of prime numbers, for example, numbers other than 1 which are divisible by themselves and 1, has occupied the lifetimes of many mathematicians. Euclid showed that the number of prime numbers is infinite; this means in turn that most combinations of prime numbers are also infinite.

All, perhaps, save one. Is there an infinite number of primes only two apart like 17 and 19, 41 and 43? Nobody knows; there is strong evidence that this is so, but no proof. Evidence that

---

* Adventures with your Pocket Calculator, *Lennart Rade and Burt A. Kaufman (Penguin)*

might satisfy a court of justice is not good enough for mathematicians.

They require an irrefutable proof. This mighty challenge, seemingly useless for all practical purposes, is one of many that now perplexes some of the most brilliant minds on Earth.

7	53	41	27	2	52	48	30
12	58	38	24	13	63	35	17
51	1	29	47	54	8	28	42
64	14	18	36	57	11	23	37
25	43	55	5	32	46	50	4
22	40	60	10	19	33	61	15
45	31	3	49	44	26	6	56
34	20	16	62	39	21	9	59

The prince of magic squares.
The sums of each column, each row and both main diagonals all add up to 260. But if every number is squared, the columns, rows and main diagonals add up to 11,180. At least 20 other whole or partial magic squares lie buried within the main square.

Luck in numbers

MANY people seem to believe in their luckiness of odd numbers. Three, seven, nine, eleven and thirteen seem to have exerted a curious mystical potency on the human mind through the ages.

Inspired by a rather muddled book on this subject , I have done some research into popular superstitions about particular numbers in the hope of discovering whether there were any good reasons for them.

Nine seems to have been the most powerfully emotive of all numbers. Being three threes, it is a trinity of trinities. The Pythagoreans concluded from this that, nine represented the deity.

Whether divine or not, nine has a most respectable pedigree. The ark of Deucalion (an ancient personification of Noah) was tossed about for nine days when stranded on Mount Parnassus. In *Paradise Lost,* the fallen angels fell for nine days when expelled from Heaven. There were nine muses, nine priestesses of the Gallic oracle, and Lars Porsena, before attacking Rome, had the prudence to swear by nine gods.

Nine appears often in folklore. That Hebrew cabbalistic spell the Abracadabra has to be pronounced nine times before it could be effective (although modern conjurors seem to neglect this requirement). People wanting to see fairies were advised to put nine grains of wheat on a four leafed clover, and nine knots tied in black wool were supposed to prevent a sprained ankle. And if any convincing detail is lacking from this catalogue, we may cite the witches in Macbeth who sang around their cauldron: *'Thrice to thine, and thrice to mine, and thrice again to make up nine,'* upon which the charm was declared to be 'wound up'.

Seven threatens to rival nine in importance. The Apocalypse mentions seven churches of Asia, seven candlesticks, seven stars, seven spirits before God's throne, seven plagues, a seven-headed monster and a seven-eyed Lamb.

Nor should we speak disrespectfully of the Seven Champions of Christendom, or of the fact that seven cities each claimed the honour of being Homer's birthplace. And we have seven seas, seven wonders of the world and seven deadly sins.

In modern culture we have *The Seven Samurai* and *The Magnificent Seven*. Seven may have secured an important victory over nine when James II lost his case against the seven bishops – for James was the ninth ruling monarch since the fall of Richard III. Nor is this all. Adding Richard's 'three' to James's 'two' plus another four because James was the fourth Stuart King, we again have the defeated nine.

What of the terrible thirteen? Thirteen is traditionally unlucky because this was the number present at the Last Supper, of whom one was a traitor. Baldur once gave a banquet of twelve in Valhalla, but Loki, the spirit of evil, gatecrashed the party making thirteen, and so Baldur was slain. Folklore informs us that it is indeed dangerous for thirteen people to be seated at table, since the first to rise will die within a year. But this danger will definitely be avoided if all the diners jump to their feet simultaneously at the end of the meal.

Numbers, like their cousins statistics, can be made to prove almost anything if one is willing to suspend disbelief. The Pythagoreans maintained that the universe consisted only of numbers, a view similar to that of some modern cosmologists who hold that everything can be reduced to geometry, to which, alas, there is no royal road.

In modern life, the most important number is ten, the basis of most mathematics and increasingly of our units of measurements. But this is only because of the accident that we have 10 fingers, which might seem a poor reason for building a shrine to a number.

One day, ten may be superseded by two, the basis of binary arithmetic and the root of the working system of computers. Forgetting, for a moment, that two was the evil principle of Pythagoras, who held that the second day of the second month of the year was sacred to Pluto and therefore unlucky, we may put two and two together (since two heads are better than one), and conclude that a statement can only be true or false: there is no third possibility.

Let us, at the eleventh hour, protest at the intrusion into our lives of these materialistic twos and tens. It may be said to have begun in 1751 when Parliament 'stole our eleven days', revising the calendar so that September 2 was immediately followed the next year by September 13.

Eleven is neglected by mystics, and perhaps deserves more attention. So far, it has attracted unfair discrimination. The Puritans were swift to put an end to the Eleven Years Tyranny of Charles I, and when the fifth century British princess St Ursula set out on a pilgrimage to Rome with 11,000 virgins they were all massacred en route by the Germans.

PLAYING GAMES

COMPUTER scientists – sorry to keep harping on them – hope to create an extraordinary new sort of indoor game within the next few years. Instead of merely watching a melodrama on television, we shall be able to take part in it, altering the story by becoming one of the characters.

But scientists are just as interested in the older games. One of them sees Sherlock Holmes battling with Professor Moriarty with the world as their chessboard.

Getting a slice of the action

Ahead, the floating arrow of light that had been their mysterious guide through the Crystal Mountain still beckoned them on. They had no choice but to follow it ... though it might lead them into yet more frightful dangers.

Arthur C. Clarke, *The City and the Stars*

THIS episode, from a novel set in the far future, is an 'adventure game' in imaginary surroundings which the characters play when they have nothing better to do.

By pressing a few switches on a machine, these people suddenly find themselves in a cavern where they must do battle with fearsome monsters. The cavern is merely an electronic projection. Yet it seems so real that the people cannot tell from their senses alone that it is all make-believe.

Like so many of Mr Clarke's ideas, it seems an accurate prediction of what will soon be happening. Home entertainment will take an astonishing new form. Instead of passively watching a television melodrama with its dialogue and story predetermined by somebody else, we may, literally, be able to become one of the characters and change the story by our action.

We can even do this today, to a primitive extent, with computer games, of which the crudest might take the form of the following words appearing on a screen:

YOU ARE THE DICTATOR OF BARATARIA, AND A HOSTILE MOB IS APPROACHING YOUR PALACE. DO YOU ...
1. Flee to Switzerland with the national funds?
2. Make a speech from the balcony?
3. Invite in a deputation?
4. Mow them down with machine guns?
Press 1, 2, 3 or 4 on the keyboard.

And after a few more questions have been answered, this message may appear:

YOU'RE USELESS! THE COMMUNISTS HAVE SEIZED BARATARIA. ANOTHER GAME?

Let us see how such simple logical games will evolve, through the

94

next two decades, into the kind of marvel that Mr Clarke describes.

The first step, in which great progress is now being made, is the steady improvement in computer graphics. Instead of merely being told of the approach of a hostile mob, we will see a picture of them. On cheap computers today, such pictures are rough and ready. A few capital Ys and Ts, with the occasional asterisk above them, will indicate a furious crowd firing guns into the air.

This is the best one can do with a screen of only 1,000 'pixels', space for a single typed character. But a high quality colour television screen contains the equivalent of half a million pixels. And a television set of the next decade may have more than double that number. One of these, when fitted to a computer, could display cartoons indistinguishable from real life.

In the words of Mr Trevor George in the April 1982 issue of *Electronics and Computing* 'films for television and the cinema as we know them could rapidly disappear. In their place will appear simulations with "actors" quite unrecognisable as the cartoon characters they are, performing stunts no real life actor could possibly achieve.'

In some areas, this is already happening. Pilots of the big jets learn to fly almost without leaving the ground. They work the controls against a computerised simulation of land and sky.

To return to that threatening mob. It seems inappropriate that they should march in silence. The machine would emit a savage roar, getting louder and louder, as 'you', the dictator, trembled with indecision. The scene would change, from the courtyard outside to the corridors within. The viewer would control the dictator's actions and even his words.

To a limited degree, machines can already recognise human speech, and advances are being made in this field almost by the month. You could stride on to that balcony, raise your fist and threaten vengeance against all traitors. The mob would then either assassinate you, flee in terror, or do something else.

Do something else? In real life, no one could foretell how the mob would react: and the machine will reproduce this uncertainty. No matter how often you will have played the game before, the mob will always seem to do something slightly different. There may be about 50 different ways in which they could react, depending on such factors as whether they are genuinely angry, or whether they are led by agitators against their will – or even how hard it is raining.

The computer program, which is the basis of the game, will randomise the consequences of your actions – just as a chess machine plays at random, with a bias towards sensible moves. Think how boring it would be otherwise. How profitless to play against an opponent who always did the same thing! The fascination of an adventure game is that nobody, not even its author, will know what is going to happen next.

Some experimenters today are devising games for several screens instead of one, so that action appears to take place all around; but with three dimensional holograms one could even improve on this. One's sitting room could vanish, and imaginary scenes would replace it. Only the lack of physical sensation will distinguish the false environment from the real.

From Aldous Huxley onwards, people have wondered how to include physical sensation. How beautiful a 'romantic' game would be if the pleasures of sleeping with an imaginary lover were not merely intellectual! But whether such 'feelies' could be created by electronics, or by any other method that society would permit, may not be known for many years.

Just as some people today are obsessed with arcade electronics, a new generation may grow up to spend most of their waking hours in bizarre 'adventures'. They may have little time for the real world, and the false, for them, may be more real than the real. Will they have the will-power to shout, like Alice: 'You're nothing but a pack of cards!' and wake from her daytime dream?

Sherlock on the chessboard

'He excelled at chess – one mark, Watson, of a scheming mind'
Conan Doyle, *The Retired Colourman*

L ORD Dunsany, the famous Irish writer, was the author of the
remarkable chess problem shown here. Its solution demands
skill in logic and no more familiarity with chess than a knowledge
of the rules. The questions to be answered are (1) why is White
overwhelmingly likely to win, and (2) however did the players get
into this extraordinary position?

Solving this kind of problem will not help a budding chess
student to beat Korchnoi, since the play has been so eccentric that

the chances of this position – or any position like it – occurring in a real game, are virtually zero. But it is a useful exercise in 'retrograde analysis', the study of the past: or, to be more exact, the study of past events from the evidence of a present situation.

Retro-analysis is therefore the root of much scientific thinking. It is as useful to the astronomer pondering the creating of the universe by observing space as it appears *now* as it is to the detective who solves a murder by deducing the series of events that led to the crime.

The solution to Lord Dunsany's problem becomes obvious when we look at the clues. It will be seen that Black's king and queen are transposed from their normal starting positions and that White has no pawns. It follows that White can attack with any of his officers, while Black cannot move any of his pawns and can therefore only move his knights.

Why? Because any movement by Black's pawns would be backward and therefore illegal. Contrary to our normal expectations, White is playing from the top of the board, and Black from the bottom!

Only if this is so could Black's king and queen have become transposed, and only thus could White have lost all his pawns. Of course, the chances that the remaining pieces would somehow arrange themselves during earlier play in this parody of the normal starting position are almost infinitesimal. But the chessboard here is being used only as a tool for an exercise in retro-analytical deduction.

A delightful book of similar 'chess problems' is available. In *The Chess Mysteries of Sherlock Holmes,* *Raymond Smullyan, a professor of logic at New York City University, gives us a series of new Sherlock Holmes adventures, told in the exciting Conan Doyle style. But there is one major difference from the great original. All the suspense comes in Holmes's retro-analysis of bizarre chess positions.

And yet there is plenty of real crime. No less than Holmes himself, the villainous Professor Moriarty, it appears, has been an expert chess retro-analyst since the age of eight.

*The Chess Mysteries of Sherlock Holmes, *Raymond Smullyan* (Hutchinson)

'On two separate occasions in my encounters with Moriarty,' Holmes recounts one day, 'I received from him a threat by mail – each in the form of a chess problem.'

'How singular!' exclaims Dr Watson.

One of these threats is indeed singular. Holmes is on the run from Moriarty, hiding himself at different London addresses each night to avoid assassination. At length a chess problem arrives at the very house where he is staying, with the message 'You certainly have a remarkable capacity for making yourself invisible, Holmes. Nevertheless, I can mate you in one move.'

The curious thing about the chess problem accompanying this message is that the White king is missing from the board. Moriarty plainly identifies himself with the Black forces, and so the 'invisible' White king must represent Holmes himself.

But what is the significance of the missing White king? Holmes with his brilliance at chess logic, quickly sees which square it must occupy to be checkmated in one move. But what is the point of the message anyway? Holmes sees it in a flash. The 64-square chessboard has been made by Moriarty to correspond to a 64-square grid map of London, and the position of the missing White king corresponds to the exact geographical location of the house where Holmes is staying. He rushes just in time from the house before it is blown up.

The episode of Colonel Marston's buried treasure is another splendid addition to the life of Sherlock Holmes. This vast treasure, a fortune once accumulated by a pirate, is hidden somewhere on a small island in the East Indies. Holmes is promised a liberal share of the treasure if he can find it, the only clue to its location being a coded message found in the old pirate's library.

Not surprisingly, the message turns out to be a series of chess problems in which a map of the island like the map of London in the earlier story corresponds to the shape of the chessboard. The greatest obstacle in finding the treasure is that Moriarty has already been there and helped himself to half of it, not knowing that it was buried in two separate places. Once again, the two masters of deductive logic seek to outwit each other by analysing past chess moves.

Despite these melodramas, Professor Smullyan's book is a powerful exposition of the way in which scientists think, and of the way in which they have to think if a difficult scientific problem is to be solved. And by making Holmes his chief character, he shows how the reasoning involved in police work and in scientific research is often essentially the same.

MATTERS OF RELIGION

WHO wrote the Book of Genesis? Why and how? Isaac Asimov suggests some interesting answers. And is there, some researchers ask, any evidence for the existence of an after-life?

An open book on the authors of Genesis

WHO wrote the Book of Genesis? Why and how? These questions are answered in a remarkable book published by Professor Isaac Asimov,* that most distinguished of science writers, who writes, not in any spirit of irreverence, but in the earnest pursuit of knowledge.

These Biblical writers and editors, say Professor Asimov, were thoughtful men who borrowed selectively from their sources, choosing what they considered good and rejecting what seemed nonsensical or unedifying. they laboured to produce something that was as reasonable and as useful as possible.

They succeeded wonderfully. There is no version of primaeval history, preceding the discoveries of modern science, that is as rational and as inspiring as the first 11 chapters of Genesis. Nevertheless, humanity does progress. Succeeding generations learn more and deduce more. If the primaeval history recounted in Genesis falls short of what science now believes to be the truth, the fault cannot lie with the Biblical writers, who did their best with the material available. If they had written what we know today, we can be certain that they would have written completely differently.

The two sources for Genesis were the J-document, written about 700BC when Assyria (modern Iraq) was the dominant power of the region; and the P-document, written much later, during the captivity of the Israelites in Babylon. The two documents have very different styles, and it becomes easy, after some practice, to tell one from the other.

The P-document is the duller of the two. It contains little dialogue and is filled with facts and figures. When you read endless accounts of so-and-so begetting so-and-so you are reading from the P-document.

---

*In the Beginning, *Isaac Asimov (New English Library)*

The J-document is much more racy. To it we owe:

1. The tale of Adam and Eve and the Serpent;
2. The tale of Cain and Abel, and of Cain's descendants;
3. Examples of wickedness before the Flood;
4. The tale of Noah and Ham;
5. The tale of Nimrod;
6. The tale of the Tower of Babel.

Even the personality of God differs greatly according to which document is speaking. In the P-document, God is an all-powerful being, creating and destroying with stern authority. But in the J-document, written when the Israelites imagined God as much less despotic and exalted, we have a more human character who seems to be the first among equals instead of an omnipotent being.

'And God said, Let us make man in our image . . .' (Authorised Version 1.26). Who is 'us'? With whom is God holding this conference? The J-document here gives the impression of a Cabinet of gods, in the polytheistic Greek fashion (although the J-document's God is a much more respectable personage than Greek Zeus).

'And they heard the voice of the Lord God, walking in the garden in the cool of the day; and Adam and his wife hid themselves amongst the trees of the garden . . . And the Lord asked, Who told thee thou wast naked? Hast thou eaten of the tree whereof I commanded thee that thou shouldest not eat?' (3.8-11).

What? Is God so human in taste that he awaits the cool evening breeze before taking a stroll? And is he so lacking in omniscience that he does not know what has passed between Eve and the Serpent before his arrival?

Throughout history, Professor Asimov explains, there have been two ways of interpreting this J-document passage. The first is that Adam and Eve hid because they were unaware of God's all-seeing powers, and that God concealed these powers, because he wanted a free confession of their guilt.

But the second explanation is more subtle. It is that God did not know of Adam and Eve's disobedience, and that he had to behave like a detective to find it out.

Now there is strong evidence that the compilers of the J-document intended to give this impression. For the Bible, produced in an age of wide illiteracy, was intended to be read aloud as a story. And a story, to be interesting, must have suspense.

In these verses, there is a great deal of suspense, whether intended or not. Will Adam give some convincing false explanation of why he and his wife are wearing clothes? Can they deny eating the fruit of the tree and be believed? Can they escape from a nasty fix? One suspects that the editors wished people hearing the story for the first time to be caught by this human interest.

Of equal interest is the tale of Noah. It appears that the compilers of the P-document lifted most of the story of the Ark from the Sumerian *Epic of Gilgamesh,* in which the hero Ut-Napishtim seeks immortality by building an Ark in which to survive a flood. Judging from their dimensions, both Noah's and Ut-Napishtim's arks were conceived by landlubbers, since they were built without keels and would have capsized at the first storm.

'And the dove came in to the Ark in the evening; and lo, in her mouth was an olive leaf.' (8. 11). This is one of the most revealing statements in Genesis, for it tells us how little the ancients knew of biology.

In fact, an olive tree which had been more than 40 days and nights under water would be dead and rotten. But the Biblical writers did not view plants as living things, but as outgrowths of the Earth. 'It would seem natural to them', says Professor Asimov, 'that once the dry land was again exposed its growth would re-form at once – or perhaps had never disappeared.'

I hope the learned professor will continue his exciting commentary and that even now he is turning his attention to Exodus.

Seeking evidence for an after-life

DOES life continue after death? This, one of the most perplexing of all mysteries, was until recently outside the domain of science, since there were no tools with which to solve it.

This situation is now changing. Scientists believe that while there is no conclusive evidence for an after-life, there is at least sufficient experimental data to make the question worth pursuing.

There is even a professional scientific organisation which is searching for evidence of an after-life. Known, most soberly, as the Association for the Scientific Study of Near-Death Phenomena, it has a membership of more than 250 which includes psychiatrists, psychologists, cardiologists and neurologists from throughout the United States.

How does one seek evidence for an after-life? Obviously, to approach the question in the manner of a spiritualist, to call up the ghosts of the dead and ask them how they were feeling, would be hopelessly unscientific. People who attempt this often find themselves the victims of hoaxes and conjuring tricks.

A more promising approach must lie in the phenomenon of near-death experience. A person is almost dead. He may come under one of the several definitions of being 'clinically dead'. Then he unexpectedly recovers, and he is later able to answer questions about his experiences.

The psychiatrist Dr Raymond A. Moody has studied more than 100 such cases, and has found much similarity in what people remember after their apparent return from death. Here is a typical experience: the patient, having apparently died, finds himself outside his body and sees it from a distance, lying prone. Then:

> After a while, he notices that he still has a 'body' but one of a very different nature from the one he has left behind. Soon others come to meet and help him. He glimpses the already-dead spirits of relatives and friends, and a loving, warm spirit of a kind he has never encountered before – a being of light – appears before him.
>
> This being asks him a question, non-verbally, to make him evaluate his life, showing him a panoramic instantaneous play-

back of its major events. At some point he finds himself approaching a sort of barrier, apparently representing the limit between earthly life and the next. Then he finds that he must return to earth, that the time for his death has not yet come . . .

But not all near-death experiences are so pleasant. In some cases studied by Dr Maurice Rawlings, a cardiologist at the Diagnostic Centre, in Chattanooga, Tennessee, people at the point of death had a decidedly 'hellish' experience. One patient saw a 'frightening human form with a goat's head'. Another reported seeing 'a huge giant with a grotesque face that was watching me. All around me I could hear people moaning.'

One cannot deny that these first-hand accounts sound highly suspicious. They have a strong whiff of religious myth. The 'loving, warm spirit' with his non-verbal question and the instantaneous playback which he provides of the patient's past life, is very like the idea which many of us have of St Peter.

Similarly, the 'goat's head' suggests a well-known mythological demon, the evil goat of Mendes recreated by Dennis Wheatley in his novel of black magic, *The Devil Rides Out*, or perhaps of H.P. Lovecraft's *Black Goat of the Woods with a Thousand Young*.

In short, while there is no suggestion that the patients were deliberately lying, it is possible that the human mind subconsciously draws upon its memories of religious teaching, taking memory for reality, as if to shield itself from the catastrophe of death.

Dr Ernst A. Robin, a neurologist at the Lafayette clinic in Detroit, points out that such delusions can be a symptom of Hypoxia, the arrest of the flow of oxygen to the brain. 'As Hypoxia persists,' he explains, 'delusions and hallucinations occur.'

But Michael Sabom, professor of cardiology at Emory University in Atlanta, sees no reason to dismiss these reported experiences as mere delusions. 'I think they are scientifically verifiable', he says. 'That what they are seeing and later recalling is something that actually did occur. I think this because of the consistency of the reports.'

What if he were right? What if mind could exist without

matter? If this possibility were admitted, then traditional physics would have to be re-written to allow the existence of thinking entities that were not subject to natural laws because they possess neither mass nor volume.

Could such a hypothesis ever be verified? It might seem that the answer must for ever be no, for there are no instruments with which to do it.

Yet this view is too pessimistic. The capabilities of science are constantly growing. A century ago, for example, it was believed that we should never know what made the sun shine. Now we have a clear explanation. In the 1920s, no one knew the age of the universe or its ultimate fate, and it was thought that these things could never be known. Today, we know roughly its age, and we have a shrewd idea of its fate.

This same march of knowledge may show us eventually whether or not there is an after-life. Already the science of quantum mechanics has indicated regions where the spirits of the dead might go, if such spirits exist. Both theory and experiments suggest the existence of other, parallel universes whose physical laws could differ from those in our own.

Whether there is a spirit world, and whether that world is within ours or within another, is a question that may well soon be removed from the domain of superstition and dogma. A solution to the mystery would, happily, make one less thing for fanatics and mystics to pontificate about. Let us hope that the American group soon makes further progress in its investigations.

PSEUDO-SCIENCE

HERE is a fascinating gallery of charlatans and maniacs. Martin Gardner, the American scourge of the pretentious and the insane, goes into battle. The more pompous of the social scientists are similarly exposed. So are those economists who make a fad of being excessively pessimistic.

But the unscrupulous hero of this section is the Great Moon Hoaxer, Richard Adam Locke.

The ways of charlatans

Mediocrities sweat blood to produce rubbish.

Anatole France

FOR every genuine scientist there are hordes of crackpots. Flat-earthers now appear to be extinct, but believers in flying saucers, the paranormal, spoon-bending, near collisions between Earth and Venus, and the literal truth of the Bible seem to abound on every side.

An eloquent scourge of these fanatics is Mr Martin Gardner, until recently the famous columnist of *Scientific American,* who has published a book called *Science: Good, Bad and Bogus.** Of good science, Mr Gardner has little to say in this volume, but of the other sort he has amassed a fascinating chamber of horrors.

Some of the worst rubbish emerges in confrontations between pure science and politics. The Nazis rejected relativity and all its works because Einstein was both Jewish and hostile to German militarism:less well known, but equally fantastic, was the violent Soviet opposition to the Uncertainty Principle – that arena of quantum mechanics which shows that the speed and position of the electron can never be determined with absolute accuracy, because the information with which to do so does not exist.

To the Soviet spokesman Andrei Zhdanov, the Uncertainty Principle was 'corrupt and vile'. 'The subterfuges of contemporary bourgeois atomic physicists,' he stormed, 'lead them to conclusions about the "freedom of the will" of electrons. Who, then, if not we – the land of victorious Marxism – are to stand at the head of this struggle against depraved and infamous bourgeois ideology? Who, then, if not we, are to deliver the shattering blows?'

What does this raving mean? What can 'bourgeois ideology' have to do with the electron? The answer, which Mr Gardner has learned from the obscure writings of certain Western Communists, is that the freakish behaviour of the electron is intolerable to a society which views the world in rigidly mechanistic terms, and that anyone who sees the behaviour of this atomic particle in this

*Science: Good, Bad and Bogus, *Martin Gardner (Prometheus Books, 10 Crescent Views, Loughton, Essex.)*

way must be a class enemy and villain, deserving to be punished with 'shattering blows'.

Against Western pseudo-scientists, Mr Gardner is himself skilled at dealing such blows. Arguing with these people tends to be a waste of time for they will not listen. To a religious fundamentalist, for instance, who is convinced that the Earth was created 6,000 years ago and was soon afterwards purged of sin by a global flood, no amount of explanation of the lack of geological evidence and of the absence of sufficient water vapour will have the slightest effect. The only answer to such people is derision. As H.L. Mencken put it, 'one horse laugh is worth ten thousand syllogisms.'

Mr Gardner belongs to the well-known Committee for the Scientific Investigation of Claims for the Paranormal, a group of scientists and other rational thinkers who make it their business to expose cranks and charlatans.

This group recently met officials of the National Broadcasting Corporation to protest at their outrageous pseudo-documentaries about the marvels of the occult. The discussion became heated, until one official shouted in a fury: 'I'll produce anything that gets high ratings!'

Would he then produce a documentary about President Kennedy's adulteries, Mr Gardner wondered? After all, they were authentic and they'd produce fantastic ratings. But the answer was probably no, for such a programme would be in bad taste. Bad taste? 'The sad fact was that not a single NBC official at our meeting knew enough about science to comprehend in the slightest the degree to which their moronic shows about the paranormal were in bad taste.'

Let us take a look at a typical pseudo-scientist. What sort of person is he, and why does he dedicate his life to producing nonsense? Such a person is likely to be an intellectual hermit. He works alone. He never discusses his ideas with people who might see objections to them.

He soon comes up with some fantastic notion, which produces an instant following of people with a taste for unlikely stories. The enthusiasm of these followers fills him with vanity. He sees himself as a second Galileo, Newton or Einstein, whose genius 'establishment scientists' are too stupid to recognise.

With vanity comes paranoia. Why does his genius go unrecognised? Is it only official stupidity, or is it something more sinister? Why, surely it is deliberate malice. The scientific establishment, terrified of the threat to itself from his superior knowledge, is plotting to suppress it. That is why no one in authority will listen to him.

The psuedo-biologist Wilhelm Reich, in a book called *Listen, Little Man*, likened himself to the persecuted figure of Christ. He then thundered: 'Whether you glorify in a mental institution, whether you adore me as your saviour or hang me as a spy, sooner or later necessity will force you to comprehend that I have discovered the laws of the living!'

The charlatan, once having started on this road, cannot turn back. He can never admit that he was wrong, even to himself, no matter how clearly this is demonstrated by others. He is, like Macbeth,

Stepp'd in so far that, should I wade no more,
Returning were as tedious as go o'er.

Sometimes a genuinely great thinker can turn crackpot. Such a person was Sir Arthur Conan Doyle. Here was one of the finest writers of the twentieth century who, in full possession of his mental powers, suddenly announced his belief in the existence of fairies.

The mind of Conan Doyle seems to have been far inferior to that of his creation, Sherlock Holmes. Indeed, it seems almost impossible to believe that the former could have invented the latter – until we observe those builders of chess-playing computers that play better chess than themselves.

So convinced was Conan Doyle of the reality of fairies and other spiritual visitors that when evidence was laid before him that he had been hoaxed, he simply refused to believe it. Not for him the penetrating intelligence of the great detective: 'These are deep waters, Watson.' The writer of these words was simply incapable of thinking like the man who uttered them.

Here, then, is something of the flavour of Mr Gardner's

splendid book. My only regret is the absence of a chapter on the lying opponents of nuclear power. In comparison with fairy-hunters and spoon benders, the damage they have done to industrial civilisation is immeasurable. Perhaps Mr Gardner, in some future essay, will expose these pestilential demagogues.

Pollution of the mind

THE most dangerous form of pollution is surely the pollution of the mind. Unlike physical pollution, it is never confined to one area: its evil is unbounded by frontiers or oceans. And of all its many kinds, the most insidious must be the circulation of falsehoods.

Two recent books, mass-produced by two well-known paper-back publishers (whom I shall not identify, for fear of making it easier for people to buy these works), must qualify as examples of the worst kind of mental poisoning.

The first is by Charles Berlitz, the principal creator of the myth of the Bermuda Triangle. This, for those fortunate enough not to have heard of it, is the proposition that an area of the Atlantic bounded by Miami, San Juan and Bermuda contains some unknown force which makes ships and aircraft disappear.

Conceivably, this may be true; but nothing that Mr Berlitz writes in support of it brings us any nearer to believing it than before. Whenever it is possible to check his facts, we find them wrong. As one authoritative reviewer noted, 'his credibility is virtually non-existent. If he were to report that a boat was red, the chances of it being some other colour are almost a certainty . . . He leaves out material that contradicts his "mystery". A real estate salesman who operated that way would end up in jail.'

The latest production from Mr Berlitz is a book called *The Philadelphia Experiment,* in which he claims that in 1943 the US Navy conducted an all-too successful experiment to make a destroyer invisible. Invisible it indeed became, says Mr Berlitz with portentous seriousness, with the result that its crewmen, after returning briefly to the world of the visible, at length became permanently invisible, and presumably unfit for duty.

One's first reaction is one of hilarity that anyone could write such stuff and expect to be believed. But Mr Berlitz's explanation of how the ship was able to become invisible does not strike me as being funny at all. For it involves a great falsehood, namely his statement that in the 1920s, Albert Einstein produced a unified field theory which would have permitted physical objects, while on the Earth's surface, to vanish into another dimension.

In fact, Einstein was never able to produce a unified field theory – the discovery of a linkage between the four fundamental forces of gravity, electromagnetism and the weak and strong nuclear forces.

And even if he had done so, there is no way that anyone could have used such knowledge to make ships disappear, The unified field theory, when and if it is completed (and successful progress is now being made), will merely show that all the natural forces of the universe are different manifestations of the same force.

It will have nothing to say about disappearances into other dimensions, since it is not concerned with such matters. Mr Berlitz might have made his story more convincing if he had opted for Einstein's General Theory of Relativity.

Am I making too much of this drivel? Perhaps not. For when millions of people (Berlitz's books are multi-best sellers) are willing to believe, without demanding proof, that Einstein constructed a theory which he did not construct, that ships disappeared that did not disappear, then it is but a short step towards believing people who say 'I have in my hand a list of 97 members of the Communist Party' or: 'The Jews are plotting against the Reich.'

The second pseudo-scientific book to provoke my indignation is *Secrets of our Spaceship Moon*, by Don Wilson.

This book sets out to prove that all the American moon landing spaceships were followed by manned alien spacecraft.

He alleges that during the first moon-walk mission in 1969, led by Neil Armstrong, the following radio conversation took place between Armstrong and Mission Control.

Armstrong: *What was it? What the hell was it? That's all I want to know.*

114

Mission Control: *What's there?* (garble). *Missin Control calling Apollo 11.*

Armstrong: *These babies were huge, sir. Enormous. Oh, God, you wouldn't believe it! I'm telling you there are other space-craft out there. Lined up on the far side of the crater edge. They are on the moon watching us.*

I myself was present at Houston during this moon walk. I took a full shorthand note of the radio conversation. My verbatim account of it, which omitted nothing of substance, appeared on page 22 of the *Daily Telegraph* on 22 July, 1969, and photo-copies are available on request. There was nothing in it remotely resembling Mr Wilson's version.

Mr Wilson says that his version was suppressed by NASA. But this is impossible! There were no silences in the conversation as it came over the loudspeaker at Houston. In short, when Mr Wilson says that a conversation about alien spaceships took place, and that NASA suppressed that conversation, he utters what Sir Winston Churchill once called a terminological inexactitude.

Crazy theories

We are all agreed that your theory is crazy. The question which divides us is whether it is crazy enough to have a chance of being correct.

Niels Bohr

HERE is the latest speculation: Jesus was an American astronaut, or rather *temperonaut* who, when time travel is invented, will go backwards in time to the Roman epoch to preach and be crucified. This journey must take place; for if it does not, then all vestiges of our religion will cease ever to have existed.

This idea, although unlikely and unsupported by evidence, is in the strictest sense irrefutable. The time journey either takes place or it does not. In neither case would anything extraordinary

115

be seen to have occurred. In the first case, the world remains as it is: and in the second, it changes but with no possibility of our detecting the change.

This notion would have made a fine short story, but unfortunately its author, a certain Mr Ronald Pokatiloff, has chosen to present it as scientific fact. 'My theory is radically different, like Darwin's was,' he says complacently of his book, *Was God a Future American Astronaut?* But in fact, unlike Darwin's, his work is a classic example of 'pseudo-science', the product of the charlatan and the crank.

This is one of many examples from an excellent book, *Science and Unreason,** which goes further than the usual exposures of bogus scientific theories. It is an essay on the nature of intellectual rubbish and it does much to explain how such rubbish comes to be written.

Here are more 'theories' which the authors have unearthed. According to the International Flat Earth Research Society, the Earth is not spherical, but flat and disc-shaped, like a gramophone record. This, says the society's journal *Flat Earth News,* accounts for its circular appearance when seen from space.

But how could mariners circumnavigate a flat Earth? That's easily answered, says the Society's founder, Mr Wilbur Voliva, who has constructed an elaborate 'alternative geography'. They simply sail round the edge of the disc. Ships do not vanish over the horizon. They just shrink with distance until they are invisible. And the earth itself is motionless. It must be so! For if it moved, would not the winds blow with continuous violence in the opposite direction to its motion?

Is this your day? According to believers in some 'biorhythm theories' there are predestined 'good days' and 'bad days' in everyone's life to determine good or ill fortune, sickness and health. One such theory describes a 23-day physical cycle, a 28-day emotional cycle and a 33-day intellectual cycle.

Birthdate and sex alone determine these cycles, which are said to be as dependable as the sunrise. Millions of pounds have been spent on biorhythm charts, guidebooks, clocks and computer

*Science and Unreason, *Daisie and Michael Radner (Wadsworth International Group, 52 Bloomsbury Street, London WC1B 3QT)*

programs. Many businessmen swear by them. Unfortunately, every penny of these millions has been thrown away. The cycles are mystical invention. In the words of one biologist: 'This is a silly numerological scheme that contradicts everything we know. There are cycles, but they vary in dozens of ways from person to person.'

People who produce this kind of nonsense, ignoring criticism of their work, rely on quotations which they often deliberately misinterpret. A certain Ray Palmer decided that another civilisation lived inside the Earth, emerging in the UFOs from holes in the Antarctic ice. He took his 'evidence' from statements by Admiral Richard Byrd, of the US Navy, before and after his famous flights over the South Pole in the 1940s and 1950s. 'I'd like to see that land beyond the South Pole,' Byrd told reporters. 'It is the centre of the great unknown.' Later he called Antarctica 'that enchanted continent in the sky, that land of everlasting mystery.' His flight was 'the most important expedition in the history of the world.'

Palmer saw these statements as proof of the hollow-earth theory. The 'unknown land beyond the pole' could not mean merely the land on the far side of the pole, since there was nothing 'unknown' about its existence. Byrd must have meant an area inside the polar regions. And the 'enchanted continent in the sky' was the surface of the Earth as seen from inside the hole.

From such 'evidence' cranks and crackpots construct their beliefs, often attracting millions of adherents. They love the quotation from the great physicist Niels Bohr, which I have given above, to show that the crazier a theory is, the more likely it is to be true. But Bohr did not mean this at all. By 'crazy' he meant extreme or unusual. It may have been a mistranslation from the original Danish. But whatever the truth of this, it has encouraged interminable folly.

Locke's 'life on the moon'

ONE of the most difficult feats to achieve in science is a successful hoax. It requires experience and talent to deceive a large number of normally sceptical people into believing in a bogus discovery, and to maintain the deception.

The Piltdown hoaxer, whoever the culprit may have been, plainly had both these qualities, constructing a relic of early man which convinced professional scientists for 41 years. So also did the New York journalist Richard Adam Locke, perpetrator in 1835 of the Great Moon hoax.

Locke's achievement was to persuade hundreds of thousands of readers of the *New York Sun* that responsible scientists had found an intelligent civilisation of winged men living on the moon.

He set about this fraud in a most businesslike fashion. Day after day, he told astonished newspaper readers that these discoveries were being made by the great astronomer Sir John Herschel, in his newly-installed observatory a few miles from Cape Town.

It was true that Herschel had an observatory near Cape Town, but nearly everything else that Locke said about him was false. Yet it sounded authentic. Locke knew enough about astronomy to write convincingly about Herschel's telescope, although in a wildly exaggerated fashion, making it ten times larger and thousands of time more powerful than it actually was.

Locke had the perfect manner which a successful conman needed to deceive both his editor and his readers. Even in appearance, he was more like the popular image of a scientist than a newspaper reporter. His face was high-domed and intellectual, his dress meticulous, his manner grave and professional.

He made no claim to have interviewed Herschel personally; much more cleverly, he stated that he was quoting Herschel's reports in the *Edinburgh Journal of Science* sent to him, allegedly, by a devoted friend of Herschel's in Scotland, a certain Dr Andrew Grant.

Might not so elaborate a story contain some truth? It contained none: but for a time, Locke got away with it.

In a world without aircraft or radio, there was no one to point out that the Edinburgh journal to which Herschel had in fact often contributed had ceased publication two years before, and that 'Dr Grant', who was thanked so politely in Locke's newspaper, did not exist.

It was two months before Herschel learned of his own fabulous lunar 'discoveries', by which time Locke himself had disappeared.

Locke wrote with a kind of high-sounding pseudo-technical phraseology that today sounds like dialogue from *Star Trek*. To explain how Sir John Herschel was able to see these extraordinary creatures jumping around on the moon (a feat that would be impossible even through modern telescopes), he described an imaginary conversation between Herschel and his real-life friend, the optical inventor Sir David Brewster.

'After a few minutes' silent thought, Sir John diffidently inquired whether it would not be possible to effect a transfusion of artificial light through the focal object of vision . . .'

In plain language, this could only mean one thing: to erect a giant searchlight with which to see the lunar creatures more clearly. But of course Locke could not afford to use plain language. To be convincing, his explanation had to be unintelligible.

Brewster agrees that this 'transfusion of artificial light' would be possible, and the two of them set to work. Soon the advanced life-forms on the moon became apparent.

They are shaped like humans about four feet tall, but with wings. Their faces *'of yellowish flesh colour, are open and intelligent in their expression, having a great expansion of forehead.'*

These beings talked, bathed in lakes, shook their wings and flew. Herschel is made to explain that *'they are doubtless innocent and happy creatures, notwithstanding that some of their amusements would but ill comport with our terrestrial notions of decorum.'*

The beginning of Locke's downfall came when he tried to be too clever. He concluded his articles with fatal reference to *'forty pages of illustrative notes, which we omit as being too mathematical for popular comprehension.'*

American astronomers had long been suspicious, and now

they pounced. A scientific deputation arrived in the offices of the *New York Sun*. Locke tried to charm them with protestations of being flattered at meeting men of such distinction, but to no avail. Where, they demanded, were these mathematical notes?

Locke feigned embarrassment. The timing of their request was really most unfortunate, for the notes were at the printers.

Which printers, they asked?

Locke gave them an address, and they set off at once. But Locke himself travelled there more swiftly, to instruct his friends at this printing works to tell the scientists, when they arrived, that the notes had been sent elsewhere.

The investigators had a fruitless quest. In the words of the scientific historian William H. Barton: 'Locke sent them on a wild-goose chase from one printer to another, all the while short-cutting through lanes and alleys to tell his friends where to direct them next. He was no impractical genius.'

When shall we see the next great scientific hoax? My hopes rose in 1978 when the Cutty Sark Scotch Whisky Company offered a prize of £1 million to anyone who could discover an alien spacecraft on earth. With the promised £1 million to spend, surely somebody could fake up some plausible-looking contraption and give the company a well-earned fleecing? But sadly it has not happened.

Yes, sadly is the right word. Scientific hoaxes are not necessarily a bad thing. They keep honest scientists on their toes.

From the faculty of fatuous research

The objectives of the research are to evaluate the wellbeing of Highland communities within a regional setting; to derive local social indicators which could eventually form the basis for a time series of key statistics on social wellbeing in such communities . . .

NO, don't look at me. I don't understand it either. The only thing clear about this dreadful prose is that it describes how the princely sum of £12,899 of taxpayers' money is being used to finance an unintelligible academic study which will probably be of little use or interest to anyone.

Now the reckless spending of £12,899 of public money will not by itself produce hyper-inflation. But a large Government agency exists to distribute such research grants, many of which seem of highly dubious value. I speak of the Social Science Research Council, which has an annual budget of approximately £20 million.

The 1981 SSRC list of research projects must be a document horrifying to those scientists who find their own more worth-while budgets being squeezed out.

Here* are some examples of its expenditures: *'Women and Religion in a Turkish Town'* (grant £25,317); *'The Development of Crying in Infancy and its Effects on the Mother'* (£35,632); *'Study of Demography in Cyprus from 1881 to the present day'* (£27,356); *'Process of Change in the East Midland Foot-wear Industry'* (£14,392); *'Car-sharing Experiments and the effect of Peak Mode on Off-Peak Travel'* (£26,366).

There is a case to be made that large-scale public spending on these kinds of studies is inherently objectionable. First, because it diverts money from serious science, and, perhaps more important, because the results are likely to be read by hardly anyone. The above passage about the Highland communities is fairly typical of the literary style of social scientists.

We have here an almost insoluble communication problem, and the reason for it is plain. When one has little of interest to say, it is hard to find an interesting way of saying it. The temptation – yielded to all too often – is to disguise meagreness of subject-matter with pompous sentences and long words. Sir Eric Ashby, in the 8 April 1982 issue of *Nature,* has picked up this gem from the Finneston report on the engineering industry:

> All those involved with manufacturing industry, whether directly or indirectly, should review their activities to ensure that they perceive and present engineering and engineers as matters of vital national concern in their own right.

**'Research supported by the Social Science Research Council, 1981' (HMSO)*

Does this sentence mean, Sir Eric asks caustically: 'The importance of engineers for industry and the nation needs to be stressed?'

For in this kind of writing, he explains, there is almost always immense trouble taken to make the reports accurate, and hardly any to make them intelligible.

But why should this be? What is there about the social sciences that seems to demand unintelligibility? The answer must be (with important exceptions, no doubt) that we are dealing here with 'soft' science, science without rules. We are asking questions that produce different answers every time we ask them. It is like a chess game where the pieces continually make illegal moves.

The physical sciences rest on theoretical laws which are supposed to be true in all circumstances; but sociology and psychology have no such laws. They cannot – for human behaviour is almost infinitely more complicated and unpredictable than that of atoms and stars.

Whether the answer is to cut the SSRC's budget by 50 per cent, I do not know. But at all events, social science should be investigated with more economy and with less excited expectations.

Otherwise, too many people will be wasting the nation's time and money by writing things like: 'It is desirable that there should be established a meaningful and responsive reciprocal personal relationship within the social group;' when what they really mean is: 'Love thy neighbour as thyself.'

When economists are false prophets

A certain economist was proud of his superior knowledge. Wishing to show off to his friends, he announced that he would make large sums by speculating in grain during a drought.

He cornered part of the American grain market in order to profit when prices rose.

But it didn't work. Farmers had contingency plans to fight drought of which he knew nothing, such as building new irrigation systems and digging new storage wells. Grain prices fell, causing him a heavy but well-deserved loss. He was baffled and furious, for the classical economic textbooks had assured him that grain prices would always rise before an expected drought. They had nothing to say about the effects of irrigation systems and storage wells.

The story is told in an excellent book about the future which is causing uproar among economists, *The Ultimate Resource,* * by Professor Julian L. Simon, himself an economist at the University of Illinois. Professor Simon's message is plain: economists often know a great deal of classical theory, but almost nothing about the real world. Hence their pessimistic long-term forecasts often turn out to be rubbish.

Their most common error, says Professor Simon, is to assume that our supply of natural resources is finite. The Earth is finite in size; therefore, they reason, the supply of its raw materials must also be finite. When these supplies are exhausted, industry must come to a halt.

This reasoning is false, explains Professor Simon, because the pessimists take the wrong definition of the word 'finite'. In mathematics it has two meanings. A 12in. piece of string is finite in length. But the string also consists of an infinite series of infinitesimal points, and in that sense it could go on for ever.

This is the only way to think about raw materials supplies. The question is not: how much of them exists? But rather: will it be economic to mine them, and if not, what substitutes will be available? Mere arithmetic is not enough. Human behaviour must also be taken into account.

A famous blunder of scarcity prediction was made in the nineteenth century by the economist W. Stanley Jevons. In 1865, he published a long, carefully-written book, *The Coal Question,* in which he prophesied that British industry would soon collapse

---

*The Ultimate Resource, *Julian L. Simon (Martin Robertson, Oxford)*

because proven coal supplies were nearly exhausted. Coal mining could no longer continue, he said, because there was not enough coal to be mined.

But what happened? Prospectors searched out new seams, hitherto considered too expensive to dig, and inventors discovered cheaper ways of getting coal to the surface. Today, coal supplies are expected to last hundreds of years if not thousands.

Jevon's error lay in considering only currently proven reserves. His work was arithmetic and nothing more. But in the real world, where arithmetic only tells part of the truth, unproven reserves, today too expensive to extract, become available when they are needed. When oil and gas run out in the North Sea, we shall no doubt take them from the offshore waters of the Antarctic.

Pessimists are fond of quoting the law of diminishing returns, which is supposed to state that we must pay more and more for less and less. But again, this 'law' is little more than arithmetic. Professor Simon points out that in the long run the opposite must apply. Supplies increase and prices fall with new scientific inventions.

To test this theory, look at the way at which proven supplies have increased since 1950. According to figures given by Professor Simon, those of oil have risen by 507 per cent, iron by 1,221 per cent, potash by 2,360 per cent, and phosphates by a fantastic 4,430 per cent.

But even if future raw materials are secure, will we not surely starve? It is a common assumption among the environmental establishment that food production is falling. It is doing nothing of the kind. Food production per head has risen 28 per cent since 1965 and more than 40 per cent since the end of the war. In fact, when large numbers of people are said to have starved, it is a consequence either of war, natural disaster, government incompetence, bad transport – or, sometimes, grossly exaggerated reporting.

The populist writer Dr Paul Ehrlich declared in his 1968 book *The Population Bomb:* 'I have yet to meet anyone familiar with the situation who thinks India will be self-sufficient in food by 1971, or ever.'

Well, Dr Ehrlich cannot have met many people familiar with

the situation, for by 1977 India was reported to be experiencing a food 'glut'. Warehouses overflowed, and the most worrying problem was the cost of storing it.

For reasons that no psychologist has yet discovered, many people love to hear bad news, even if it is false. Books, articles and television programmes which predict disaster attract far more attention than those which assure us that all is well. For this reason, more of them are produced. But however many times gloomy prophecies are repeated, it does not make them true.

Why is man such a pessimistic species? The answer must surely lie in our long history in the primaeval forest, when we had no science to assist us, and when mortal danger was a daily threat. The achievements of the past three centuries will not easily remove the conditioning of millions of years. More voices like Professor Simon's must be heard if our intellectuals are to be aroused from their usual mood of panic and gloom.

And now for the good news . . .

HOW badly – or well – is the world going? People who rely for their information on certain books, magazines and television programmes will assume the worst; that hundreds of thousands of people are dying every year from drought and starvation, that population growth is fast out-running food production, that arable land throughout the tropics is turning into deserts, and that hunger and misery dominate much of the world.

While it is true that the so-called Third World is undoubtedly much poorer and more vulnerable to natural disasters than the industrialised countries, many of the doom-laden stories circulated about it are either much exaggerated or even plain falsehoods uttered by officials of international agencies, who presumably hope to increase their funding.

Let us take a few examples of what Professor Julian L. Simon, whom I quoted in the last chapter, calls 'an over-supply of false bad news'.*

A few years ago *Newsweek* magazine reported that 'more than 100,000 people perished from hunger because of drought in the Sahel region of West Africa between 1968 and 1973.'

The magazine took this figure from an emotional speech by the Secretary General of the United Nations, Dr Kurt Waldheim. But where did Waldheim get his information? His own source was a one-page memorandum to the United Nations by a lady who specialised in West African demography. But she did not authenticate the deaths of 100,000 people during this five year period! The part of her report on which Waldheim based his statement read as follows:

'The highest death-rate in any group of nomads during the drought was an absolute and most improbable, upper limit of 100,000. Even as a maximum this estimate represents an unreal limit.' But Waldheim from this statement alone baldly announced that the actual number of deaths was 'more than 100,000'.

In the Autumn of 1977 the *New York Times* ran a front page story headlined: *'14 Million Acres a Year Vanishing as Deserts Spread Across the Globe'.* The figure was based on an estimate published by the Worldwatch Institute. But the truth is exactly the opposite. In a survey of 87 countries, making up 73 per cent of the land area of the world, Dr Joginder Kumar, of the University of California, found that the total amount of arable land had actually increased by 9 per cent between 1950 and 1960.

I apologise for all these statistics, but without them the argument would be unintelligible. In the 1960s deserts themselves continued to vanish at an encouraging rate. The United Nations Food and Agriculture Organisation reported in 1974 that the total amount of 'arable and permanent cropland' was increasing by an average of 1 per cent a year. So much for the much-touted threat of 'desertification'.

---

*Professor Simon's article on the 'Over-supply of false bad news' appeared in Science, 27 June 1980

Consider the 'threat' of overpopulation. In theory, it could become a threat. There are now an estimated 4,200 million people in the world, and this number is growing by about 1.6 per cent a year. If this rate of increase was maintained indefinitely, a century from now the world population would be 20,000 million; and at the end of 1,000 years it would be 32,000 trillion.

Obviously, such predictions are absurd since there would never be enough food to maintain so many people on one planet. The rate of increase will therefore continue to drop during the coming decades, as it has been doing since about 1970.

Organisations like the World Bank and the Agency for International Development hold it is an almost gospel truth that higher population growth implies lower per capita economic growth. Their reasoning goes like this: every child born means one more mouth to feed.

But this argument is nonsense, since it ignores the fact that while children consume food without producing it, they have the habit of growing into adults, who both produce and consume, and the impact of their existence per capita on economic growth is then reversed.

International bureaucrats do not understand this. They maintain that, because these children cannot produce anything in the short term, their influence on the world economy must therefore always be negative.

One need not be a genius to see the error. Our short term is somebody else's long term. In other words, the children who were born more than 20 years ago – exciting alarm in the pessimists at the time – are now becoming productive. Therefore, up to a certain physical limit, and barring the effects of wars, the human race will always have the capacity to create enough food for itself.

Our descendants face a future of unlimited opportunity. Yet we have to recognise the psychological danger, the damage to the spirit of optimism, which is done by the purveyors of false bad news.

Futurists of the sillier kind

IMAGINE a hall filled with scientists. They are good humoured and scrupulously polite to each other. If a toe is trodden on, apologies are profuse. When someone spots an acquaintance, there are cordial waves and handshakes. When the lunch break is announced, they sit down together and chatter merrily, like friends who have looked forward to this moment for ages.

What are they doing, these amiable ladies and gentlemen? They are waiting for their turn to go to the rostrum and denounce mankind as the most ferocious, bloodthirsty, unco-operative and obstinate species who ever lived on this planet, and whose tenure on it is most surely doomed.

It is a typical international 'scientific' conference about the human predicament. Where and when it was held does not greatly matter. There are dozens of such conferences every year, and their tone was aptly satirised in Arthur Koestler's 1972 novel *The Call Girls*.

They are generally attended by scientists who are far too specialised and academic in their outlook to have much know-ledge of the everyday world. This is an account of one such junket which I went to, along with some of my reactions.

Animals, of which humans are but one type, cannot endure one another's close proximity, speakers assure the audience. They declare that the law of the Territorial Imperative predicts violence when one group encroaches on the domain of another. A human being becomes anxiety-ridden when others are too close to him. There is warm agreement from the scientists present, all of whom are sitting shoulder to shoulder.

One speaker chills the blood with his talk of over-population. People, he cries, will multiply until they outnumber the stars. Their annual rate of increase will itself increase. (The fact that it is actually dropping is dismissed as a statistical hiccup.) Schemes of birth control, he goes on, cannot defeat the sexual urge. Only starvation and war, he opines, can prevent the Earth from being overloaded with such a mass of humanity that its orbit round the sun will begin to be affected.

At this, there is strident applause from the middle-aged

128

audience, of whom one quarter, I discover from a straw poll, have no children, and of whom a half have fewer than three.

Ever-rising prices of consumer goods will destroy civilisation, another orator declaims. Everything is becoming more expensive. The extent of this fearful trend he can demonstrate – by pushing buttons on his £5 pocket calculator that a few years before would have cost more than £100.

Pollution of the air, water and seas is getting worse, says another speaker, drinking deeply from a carafe in front of him – which everyone knows has been filled from a tap. Pausing in his address, as if seeking inspiration, he glances through the window at a distant church spire, that 20 years ago, before the enactment of clean air laws, would have been obscured by industrial smog.

The resources of the earth are finite, the audience is told, and within little more than a century they will be exhausted. An economist demonstrates this 'fact' by displaying a nickel coin which cost more to produce than it is worth.

Ah, but where did the nickel come from? I inquire. Originally, from the famous nickel mine at Sudbury, Ontario. And where did the mine itself come from? From the bowels of the earth? Well, not exactly. The mine is the remains of a gigantic nickel and iron meteorite which crashed into the earth several million years ago.

Where, in turn, did the meteorite come from? Was it a stray missile from some unknown region in space? No, it was an asteroid, a few hundred yards in diameter, of which there may be a thousand million now in the Asteroid Belt that lies between the orbits of Mars and Jupiter.

It is estimated that the mineral resources of the Asteroid Belt will provide the space industries of future generations with the raw materials for almost any substance that man could desire. Even one such asteroid might contain enough gold to bring the world price down to $5 an ounce. So much for the accurate but misleading phrase 'the finite resources of the earth'.

It is time to go home. The delegates, still grumbling about the deteriorating environment, complain at the airport about a 30-minute delay in take-off – little reflecting that their journey will be 30 times faster than any expedition of their great-grandfathers and that the first voyage of Columbus was delayed, not by 30 minutes, but by 18 years.

Once airborne, they may dine on an excellent course of fish caught in the Mediterranean, that sea which several of them had described at the conference as having been poisoned by industrial wastes and consequently 'as stagnant as the tarn of the House of Usher'.

There is too much pessimism today among professional 'futurists'. It is perhaps that scholars, however brilliant in their specialised fields, are too obsessed with theory, and think it 'unacademic' to seek explanations of the world outside their narrow intellectual domains.

Keep it complicated

'It must be wonderful, for I cannot understand it.'

Beaumarchais

SCIENTISTS, perhaps the most far-seeing of people, are sometimes the worst writers. All too often, they delight in using jargon that renders their work incomprehensible to outsiders. Many of them love to use the most arcane words and complicated sentences to make it seem more important. In short, they indulge in Unintelligible Profundity, which I abbreviate as UPR.

UPR is especially encouraged by those editors of scientific journals, who seem to consider an 'authoritative appearance' more important than the communication of knowledge. The contagion spreads to university science departments. 'Don't use such simple words. Make it sound more scientific,' a supervisor is likely to tell a researcher after seeing a first draft of his paper.

The final draft is accordingly filled with UPR so that a scientific journal will accept it for publication. Many scientists have taken their time to learn this lesson. They make the mistake of simplifying the language of their papers, which are then promptly rejected. They then submit the original draft, full of verbose

130

phrases, which is thereupon accepted by the very same editor who turned down the better version!

The British weekly journal *Nature* is heavily laced with UPR, although I have no evidence that its inclusion is deliberate. Dr David B. Jack, a Birmingham University scientist who, with his colleague Martin Gregg, is making a spare-time study of UPR, has told me 'I usually only understand about a tenth of what I read in *Nature*.'

I have invented the following gobbledegook to show how a typical Nature article might begin. It has no meaning at all, but it is an accurate satire of the high scientific style:

The G-refractory subjective index (GRSI) is shown by us to derive from that of Murchison and Weber by several orders of magnitude. The immediate effect of the demonstration is to render nugatory any hypothesis that dextro-rotational ice crystal could have created the chordal chasm.

Authors of this kind of prose seem to go to extraordinary lengths to prevent anyone outside their field of expertise from discovering what they are talking about. It is as if they dreaded seeing a newspaper headline the next day: CHASM RIDDLE BAFFLES BOFFINS. After all, scientists are not supposed to be 'baffled'. They prefer to cultivate the impression that they alone understand the deeper mysteries.

Social scientists are probably the most prolific in their use of UPR. This is because, unlike the traditional sciences, which often deal with obscure ideas, sociology is about the ordinary affairs of ordinary people. Sociologists therefore feel under pressure to use the maximum amount of UPR, so that no one suspects them of uttering commonplace platitudes.

Suppose they were writing about a slum. A slum? Such a simple word would never do! They must talk instead of a depressed socioeconomic area. Stupid children are called juvenile underachievers. A salesman becomes a marketing representative. Death becomes a morbid terminal event, and these phrases in turn lead to such lofty concepts as adjustment alternatives, situational interactors, and coherent social consciousness.

A devasting blow to the reputation of the social scientists

occurred in the United States in 1972 with the 'Dr Fox' experiment. Some 50 senior educationists, administrators and sociologists were invited to hear a lecture by a person called Dr Myron T. Fox on the subject of *'Mathematical Game theory as applied to the Education of Physicians'*.

But 'Dr Fox' was a professional actor. He knew nothing of his subject. His lecture, for which he had been carefully coached, consisted of double talk, meaningless words, irrelevant humour, false logic, non-sequiturs and references to unrelated topics. Nevertheless, according to a questionnaire circulated afterwards, the audience found the lecture clear and stimulating, and not one of them realised it was pure nonsense!

You too, can excel in UPR. Mr Steve Aaronson, a distinguished student of the subject, has compiled a list of well-known literary quotations, with alternative and more 'impressive' versions. I have numbered them 1 and 2 in each case:

1. *I am no orator as Brutus is.*

2. I am not what might be termed an adept in the profession of public speaking, as might probably be stated of Mr Brutus.

1. *Render unto Caesar the things that are Caesar's.*

2. In the case of Caesar, it might be considered appropriate, from a moral or ethical point of view, to render to that potentate all of those goods and materials of whatever character or quality which can be shown to have had their original source in any portion of the domain of the latter.

1. *God said: Let there be light! And there was light.*

2. God, in the magnificent fullness of creative energy, exclaimed: Let there be light! And lo! The agitating fiat immediately went forth, and thus in one indivisible moment the whole universe was illumined.

1. *I returned and saw under the sun, that the race is not yet to the swift, nor the battle to the strong, neither yet bread to the wise, nor yet riches to men of understanding, nor yet favour to men of skill, but time and chance happeneth to them all.*

2. Objective consideration of contemporary phenomona compels the conclusion that success or failure in competitive activities exhibits no tendency to be commensurate with innate capacity, but that a considerable element of the unpredictable must invariably be taken into account.

Unnatural 'Nature'

WHY do so many people decline to take any interest in science or engineering? The usual answer, that they find these subjects hard to understand because they tend to be written about in difficult jargon, may be misleading. The real objection, it seems to me, is not so much to the jargon but to the vile English that accompanies it.

It is hard enough having to wade through a text that contains such unfamiliar words as 'maser', 'pulsar' and 'metagalaxy'. But it becomes ten times harder when such words and phrases are abbreviated into groups of initials and dots; when they are strung into sentences of intolerable length, not even broken by colons or semi-colons, and when the main verb has either been omitted or has been cunningly hidden in a subordinate clause.

Where, then, should we seek examples of bad scientific writing? Perhaps most of all in *Nature,* the British journal that was founded 114 years ago with the stated purpose of making science intelligible to ordinary people. Today, it does the opposite. Not only do most of the articles in *Nature* seem to non-scientific people as if they were written in a foreign language, but they are often incomprehensible to scientists in the field under discussion.

Since *Nature,* although privately owned, is widely regarded as the 'official' organ of British science, I have taken the liberty of picking from recent issues sentences which might have given Fowler apoplexy. The italics are my own:

It has become generally accepted that superluminal motion is observed in the nuclei of some radio sources.

133

Why not simply say: 'Certain objects in the centres of exploding galaxies appear to travel faster than light'? It would attract much wider interest.

An inspirational 1972 paper by Niles Eldredge and Stephen Gould has provoked fruitful empirical work and active debate about whether evolution proceeds mostly by phyletic gradualism or by punctured equilibrium (Stanley's 'rectangular' evolution).

My translation: 'People are wondering whether evolution proceeds gradually or by sudden jumps.' The adjectives 'inspirational', 'fruitful' and 'empirical' are clearly redundant. And this is an example of the confusion, typical among some scientists, of trying to cram too much information into one sentence. The references to Eldredge and Gould should have been postponed until later. And there should have been a sentence of its own to explain what Stanley means by calling evolution 'rectangular.'

It is a direct consequence of the Einstein equivalence principle (EEP) that all atomic clocks will run at the same rate if situated at the same point in spcae-time.

We hardly need Einstein to tell us that two accurate clocks in the same room will run at the same speed.

The Arabian Peninsula often appears totally aseismic on maps of world seismicity, reflecting the apparent absence of earthquakes during the 20th century and particularly since the improvement of the worldwide seismological network after the 1960s.

Why not just say: 'The Arabian Peninsula appears to have been free of earthquakes'?

For at least the past 10 years, a mild controversy has simmered in meteoritical and astrophysical circles regarding the strange distribution of xenon isotopes found in certain meteorites.

Having rewritten this to say: 'People have been wondering for 10 years in an idle sort of way why some meteorites contain xenon,' one must severely criticise it. Who else but those in 'meteoritical and astrophysical circles' would be arguing about meteorites. Linguists? Sanitary engineers? Surely not. It is unnecessary verbiage. And the statement that the controversy is 'mild' is sufficient to kill any interest in the subsequent discussion.

Measurements of stratospherical minor constituent gas concentrations are important for the elucidation of the complex chemistry amid the evaluation of the predictions presented by theoretical models regarding the short and long term future of this important atmospheric region.

I would have said: 'To understand the stratosphere one should study its gases.'

The reappearance of Van Nostrand's Scientific Encyclopedia *in a new edition (the sixth) only seven years after the last edition, combined with its sheer bulk (3,067 pages, 700 more than the fifth edition), provides welcome assurance that this very considerable work is now a permanent feature of the scientific landscape ... the Mont Blanc, if not the Everest (the 15-volume McGraw Hill* Encyclopedia of Science and Technology *is surely that), of scientific encyclopedias.*

I think this run-amok sentence means: 'Van Nostrand's *Scientific Encyclopedia* is nearly as good as McGraw Hill's, but it's hard to be absolutely sure.

Some people would argue, in perfect seriousness, that contributors to *Nature* must write in this convoluted way to preserve its editorial character. Even if very few understand it, they say, it sounds magisterially impressive, and that is what matters.

The argument does not seem very sensible. *Nature* has a weekly circulation of nearly 26,000, while comparable American scientific journals have circulations of nearly a million – because they are so much better written. As a result, most Americans seem to know much more about science than most Britons.

'The A-bomb will never go off' and I speak as an expert in explosives

The film is apparently meaningless, but if it has any meaning it is doubtless objectionable.

British Board of Film Censors in
1956 banning Jean Cocteau's film
The Seashell and the Clergyman

THROUGHOUT history, prominent people and organisations, like the one above, have been making idiotic statements, inaccurate predictions and wildly misguided criticisms. A splendid collection of these mis-sayings has been gathered into book form.* Taken together, they give credence to the suspicion of a fundamental weakness in the logic circuits of the human brain.

Cocteau was not the only artist – or author – of eminent talent to be savaged by people who might have known better. The poetry of Keats was dismissed by Thomas Carlyle as *'fricassee of dead dog',* and the works of James Joyce were ill received by his rival D..H. Lawrence.

'My God,' wrote Lawrence 'what a clumsy *olla putrida* Joyce is! Nothing but old fags and cabbage stumps of quotations from the Bible and the rest, stewed in the juice of deliberate, journalistic dirty-mindedness.'

Even Shakespeare has not escaped abuse. 'With the single exception of Homer,' said Bernard Shaw, 'there is no eminent writer, not even Sir Walter Scott, whom I can despise so entirely as I despise Shakespeare when I measure my mind against his. It would positively be a relief to dig him up and throw stones at him.'

With famous authors receiving such abuse, one might turn from them to television, but this could be difficult; for according to some experts it does not exist. 'Television won't matter in your lifetime or mine,' said Rex Lambert in an editorial in the *Listener* in 1936. He was echoed twelve years later by the famous educational radio broadcaster Mary Somerville. 'Television won't last,' she declared. 'It's a flash in the pan.'

Television is not the only modern contrivance that, according to the experts, does not really exist. Nuclear weapons? There are no such things. 'The atomic bomb will never go off, and I speak as an expert in explosives,' Admiral William Leahy told President Truman in 1945. The admiral was in good company. 'Atomic energy might be as good as our present day explosives, but it is unlikely to produce anything very much more dangerous,' said Winston Churchill in 1939.

Long distance travel must be very difficult indeed, for the words of Lord Haldane, Secretary for War in 1907: 'The aeroplane will never fly.' At about this time, the chief of British Customs ordered his officers to ignore people who arrived in the country by air, for to do otherwise 'would only bring the department into ridicule.'

Well, if aircraft will not fly, or only become ridiculous if they do, should we not travel by sea? It is out of the question. 'Men might as well expect to walk on the moon as cross the Atlantic in one of those steamships,' declared the eminent nineteenth-century scientist Professor Dionysius Lardner. Unable to travel, we cannot even send messages to distant lands. 'Radio has no future,' declared the great Victorian physicist Lord Kelvin, who also prophesied that 'X-rays will prove to be a hoax.'

The female sex has been the subject of many strange observations. Tolstoy remarked flatly that 'in the highest society, as in the lowest, woman is merely an instrument of pleasure'; and the sociologist Orson Fowler, in his 1870 book *Sexual Science,* enlarged on this view with academic exactitude: 'Every well-sexed woman invariably throws her shoulders back and breasts forward as if she would render them conspicuous, and further signifies sensuality by a definite rolling motion of her posterior.'

Judges seem peculiarly dotty on this subject. 'Athletic competition builds character in our boys. We do not need that kind of character in our girls,' said an American judge in 1973, ruling against an athlete who was excluded from a school team on the ground of her sex. Or perhaps judges have a tendency to say odd things. 'All Berkshire women are very silly. I don't know why women in Berkshire are more silly than anywhere else,' said a judge at Reading County Court in 1972.

Science writers sometimes utter the most hilarious inaccuracies. Frank Ross, in his 1956 book *Space Ships and Space*

Travel, had this to say about cosmic rays: 'Their speed is incredible, approximately that at which light travels, 186,000 miles per second. This is twice as fast as the speed of a bullet leaving the muzzle of a U.S. Army rifle.' Twice as fast? Well, no, actually. About a million times faster.

This might have been a misprint – which reminds us of many weird misprints in well-known books. The Bible has been a particular victim, usually from the omission of the words 'no' and 'not'.

'Thou shalt commit adultery,' said the edition of 1631. 'The fool hath said in his heart there is a God . . . Know ye not that the unrighteous shall inherit the Kingdom of God,' said the edition of 1653. And in 1702, the Bible itself seemed to be answering back: 'printers (instead of princes) have persecuted me without a cause' – and the printers paid for these errors with heavy fines and threats of the pillory.

It makes little odds if people are punished for misquoting divine scripture, for divines produce quite enough nonsense on their own account. Only recently, the mullahs of Iran instructed their aircrews to bomb a US reconnaissance satellite; perhaps faith may move mountains, but it cannot yet achieve orbital velocity.

It goes without saying that natural disasters are a divine punishment for sin. A fierce Boston Puritan in 1755 blamed the earthquakes of that year on the lightning conductors newly invented by Benjamin Franklin, for to avoid lightning was to hide from God's justice.

And some clergymen themselves have a strange view of sin. We have it from Bertrand Russell that one eminently orthodox Catholic divine laid it down that a confessor may fondle a nun's breasts, 'provided he does it without evil intent'. But tolerance in one field does not necessarily lead to tolerance in another. The great Methodist leader, John Wesley, once warned 'The giving up of (a belief in) witchcraft is in effect the giving up of the Bible.'

Perhaps religion may be coming to an end anyway. The late John Lennon said in 1966, 'Christianity will go. We're more popular than Jesus now.' Wait around a hundred years, and we'll see if he was right.

ANIMALS

NEANDERTHAL men – half human, half animal – may still be among us. Some scientists, with a general lack of success, have been trying to teach animals to understand human speech, and others have been employing them to teach us economics. Our own instincts of ferocity and intelligence can be traced to 'layers' built into the brains of our ancestors millions of years ago, and some attention is drawn to the noble elephant, which may be fast becoming extinct.

The brains of better heads . . .

IN a laboratory in Wichita, Kansas, there is an old cardboard box that for some reason bears the label 'Costa Cider'. Inside is a jar of formaldehyde. And in the jar is all that remains of what was perhaps the most powerful and versatile machine that ever existed – the brain of Albert Einstein.

Einstein left his brain to science when he died in 1955, and numerous experiments have been made on it to discover what made it different from other people's. The result? Nothing. The grey mass appears indistinguishable from the brain of any ordinary person.

The brains of many other remarkable people have been examined after their deaths, and always with the same inconclusive results. Indeed, some of them have been so small, no larger than a chimpanzee's, as to give no hint of the mighty intellects they housed. Here are some of their weights: Anatole France, 2lb 4ozs; Walt Whitman, 2lb 15ozs; Daniel Webster, 4lb; Bismarck, 4lb; Oliver Cromwell, 5lb; Lord Byron, 5lb; Victor Hugo, 5lb.

Plainly, it is not the size of the brain that matters, but rather the internal organisation of the mind – of which all trace appears to vanish at the moment of death. Yet certain definite facts have been discovered about the construction of the human brain, a construction that began some 300 million years ago, when the first reptiles appeared on earth.

Our brains are in fact three brains in one, each being a fresh layer wrapped around the last, and each essential to our existence.

Deep down and very small in comparison with the other layers, is the reptile brain, that biologists call the hypothalamus, which makes us capable of fighting, hunting, killing and of outbursts of rage and hatred. This we have inherited from the ancestral reptile, who was not only the ancestor of crocodiles, snakes, lizards and birds, but also of all the apes, including man.

The ancestral reptile was of low intelligence, but it knew how to survive. Fight or flee; these were the sole choices of action in which it was interested. It had no parental instincts, and it devoured as many of its young as it could catch. It would not have appreciated the Sermon on the Mount. It did not love its enemies, nor did it turn the other cheek when smitten. We have inherited

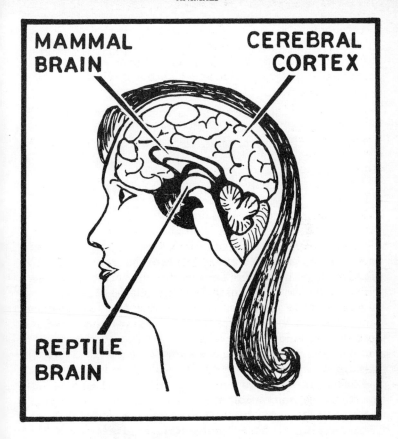

the entire brain of the unattractive but efficient beast, and it is as well that we have; for a species that allowed Christ's precepts to drive out the fighting impulse would soon become extinct.

Two hundred million years later (100 million years ago), one descendant of the ancestral reptile was forced by circumstances to develop a new and more subtle brain. The ancestral mammal, little larger than a squirrel, was in constant danger from predators. To defend itself, it developed a sense of smell and a large memory for shapes and appearances. The American scientist Robert

141

Jastrow, in his fascinating book *The Enchanted Loom,* *puts it like this:

> The mammal on the forest floor has seen many trees in its lifetime; it has learned how the appearance of a tree changes with angle, light and distance. Drawing on this stored knowledge, its brain quickly makes calculations about the appearance of a particular tree and its leafy canopy. It compares them with the unfolding evidence of its eyes; if the calculation agrees with that evidence, the brain flashes a reassuring signal to the mind: *'This object is the same safe tree that appeared on the eye's retina a few moments ago. Relax. Continue feeding.'* Or it flashes an alarm: *'The object, viewed from a new angle, does not agree with the predicted form. A predator is concealed in those branches. Take flight!'*

Nearly another 100 million years went by, and a still more powerful brain appeared, enveloping the other two. It was in the habit of asking itself questions about trees which would never have occurred to the mammal brain.

What kind of wood are they made of? Are their branches suitable for making clubs, spears and bows? Can their trunks be hollowed out and used as boats? And what of those reeds – ancestors of those self-same reeds where the ancestral mammal was wont to fear prowling death? Why, lashed together they could be made into a raft that could carry its designer across the trackless oceans in pursuit of a vision or a dream.

This new brain was the cerebral cortex, the source of creativity, science, art, and all those abilities that make man almost god-like among the animals. It appeared with explosive suddenness in the course of the past million years. It made tools, and existence of the tools in turn made it yet more intelligent. In the language of electronics, it became *'packed with circuitry'.* In the merest fraction of a geological time, a nomadic species that hunted mammoths turned its hand to skyscrapers and spaceships.

When a nation goes to war, the three brains act in formidable harmony. The reptile brain fights with the maximum ferocity.

*The Enchanted Loom, *Robert Jastrow (Simon and Schuster, New York)*

The mammal brain concerns itself with munitions, supply and all matters of routine. And the cerebral cortex invents cunning strategy and new weapons.

What of the future? Is there to be a fourth brain mightier than the other three? If so, it will be an external aid rather than internal tissue. In the past, our tools gave us more intelligence, and now we seek to give more intelligence to our tools. Super intelligent machines may succeed man, and if we are lucky, says the computer scientist Marvin Minsky, 'they may decide to keep us as pets'.

The cavemen among us

Presently came other glimpses of lurking semi-human shapes and grey forms that ran in the twilight.

H.G. Wells, *The Grisly Folk*

ONE of the greatest mysteries is the fate of early man, the brutish Neanderthals who were the direct predecessors of modern man and who populated Europe and Asia some 50,000 years ago. How did they become extinct? Some people believe they simply died out, others that our ancestors destroyed them. But the most startling recent theory is that none of these things happened; that they never became extinct, and that they are still alive.

At the back of human consciousness through the ages there has been the feeling that we are not alone, that we share the planet with unseen, but real creatures of human shape but not altogether of human form. This idea is responsible, perhaps, both for the thrill obtained from ghost stories and the accounts by the writers of classical times of such beings as satyrs, centaurs and shaggy gorgons said to be agile enough to strangle leopards.

Could such creatures be real, as claimed by such ancient authorities as Diodorus and Plutarch? And if they were real, what could they be?

143

The most likely explanation is that they are survivors of Neanderthal man, so called after the misshaped prehistoric skeleton found in 1856 in a cave of the Neander valley near Dusseldorf, and later recognised as humanity's nearest cousin, apparently long extinct.

But many scientists today do not accept the assumption that they became extinct, in particular Dr Myra Shackley, lecturer in archaeology at Leicester University, whose investigations of these beings appears in the March 1982 issue of the journal *Antiquity*. 'The idea that they must be extinct', she explains, 'on the grounds that modern man can be the only surviving hominid species, is out-moded biological arrogance.'

Where, then, are the modern Neanderthals? Apparently in remote parts of Russia, in the northern Himalayas and in Siberia and Mongolia, where there have been numerous sightings of 'wild men'.

They avoid all contact with modern man and live in the wilderness. According to those who have seen them, they have 'protruding brows and are without properly developed speech, communicating only in modulated whistles.'

A more detailed description was given in the 1900s by the St Petersburg zoologist and explorer V.A. Khaklov. He saw some of the 'wild men' and gave this description: 'They are of medium height, with hair all over their bodies. They have no forehead, but prominent brow-ridges and heavy lower jaw and no chin; with long arms and short legs. The feet are broad with the big toe shorter than the others, but massive and projecting inwards, with the other toes fanned out.'

Dr Shackley comments that this sounds very much like a classic anthropologist's description of a prehistoric Neanderthal. So also was that of the Bolshevik general, Mikail Topilsky, who encountered one when pursuing White Russian forces through the Pamir mountains of southern Russia in 1917. His troops found strange footprints in the snow near a high cliff. These led to a cave occupied by an unfortunate 'wild man' who, on emerging from it, was promptly shot.

Topilsky had the creature examined by a doctor. What was remarkable was that although it was thickly covered by hair,

144

almost like a bear, it was definitely human. 'The teeth', he said, 'were large and even and shaped exactly like human teeth. The forehead was slanting, and the eyebrows were very powerful. The protruding jawbone made the face resemble the Mongol type of face. The lower jaws were very flat and very massive.'

These do not seem to have been any wild travellers' tales, for similar reports had been appearing in the writings of explorers for centuries. Hans Schiltberger's *Journey into Heathen Parts,* published in Germany in about 1430, contains this extraordinary story about a Tartar prince named Tschekra and his companion Egidi:

> Tschekra (says the German account) joined Egidi on his expedition to Siberia, which it took them two months to reach. In that country there is a range of mountains called Arbus (probably Tien Shan) which is 32 days' journey long. The inhabitants say that beyond the mountains is the beginning of a wasteland which lies at the end of the earth. No one can survive there because the desert is populated by so many snakes and tigers. In the mountains themselves live wild people, who have nothing in common with other human beings.
>
> A pelt of hair covers their entire bodies. Only their hands and faces are free of it. They run around in the hills like animals, eating foliage and grass and whatever else they can find. The lord of the territory made Egidi a present of a couple of these forest people, a man and a woman.

Now sensationalist literature is full of accounts of extraordinary beings, from bug-eyed visitors from other planets who arrive in flying saucers to monsters like the Abominable Snowman and the American Bigfoot. It is probably safe to say that all these stories are the result either of sighting errors, hoaxes or lies.

The Neanderthal stories are in a different category. There is nothing impossible about them. We can be sure that the denizens of flying saucers would find it difficult to land on Earth without the whole world finding out very quickly about their visit. Bigfoot and the Abominable Snowman could not have existed so long without being photographed or shot. And the fact that neither of these things has happened seems strongly to suggest that such creatures do not exist.

But the supposed Neanderthals have been seen by a large number of people. Herdsmen in Outer Mongolia even testify to having shared their camp fires with them. At the very least, all this evidence implies that recent human evolution has been more complex than anthropologists have tended to believe.

History marches on itching feet

Heave away, heave away, heave away, heave away, —
For we're bound for Australia!

Old sea shanty

MIGRATION, the willingness to make long journeys into the unknown to find a better place to live, even if it means braving the hazards of storm, privation and piracy, has been for countless thousands of years a dominating urge of mankind.

So deep-rooted is this phenomenon that it becomes astonishing that anyone should believe that today's national boundaries are somehow fixed for eternity. it seems certain that a map of the world drawn a thousand years hence would seem as bizarre to us as today's would apear to a geographer of 400 years ago.

Consider the world in 1600. The United States, Canada and Australia did not exist. Germany was a mass of petty princedoms. Italy was little more than a collection of independent city-states. Russia was the small kingdom of Muscovy, of little importance in European affairs. Africa was mostly unexplored and Japan, which may soon dominate the world's economy, had barely been heard of. It is the force of migration, perhaps even more than that of war, which has changed the world of 1600 into the world of 1980. Dutchmen, despising the easy-going religion of Holland, set out to build a new and more austere nation in South Africa. Puritans in England, hating the Stuart kings, created the beginning of the United States. Russian conquerors swept relentlessly eastwards, bringing innumerable Asiatic peoples beneath their sway.

Why do people migrate? Why, for that matter, do animals migrate also? The question is partly answered in a beautifully-illustrated and scholarly book, *The Mystery of Migration.** Salmon, whales, birds of many kinds, all migrate according to the season or to various time-cycles but – and here is the astonishing part of it – their routes and times of migration can be quite unpredictable.

In a large number of animal species, reports Dr Baker and his team, there are irregular periods of restlessness which leads to 'removal migration', from which there may be no return. The group sends out explorers, like Moses dispatching spies into the Promised Land. If these are successful, the whole colony follows.

This phenomenon can be seen throughout nature. Winged kings and queens of ants and termites fly away from the parent colony in search of a new site. Crabs and lobsters are continually acquiring new breeding areas.

The spores of plants go where the wind carries them, creating new colonies. So also do many insects, such as greenflies and small beetles, which are such weak fliers that they cannot prevent themselves from being blown downwind whenever they take to the air. Unlike those migratory birds which regularly spend winter and summer in the same two countries, these insects normally migrate by flight only once in their lifetimes.

From this confused picture a general rule can be constructed. Life, whether human, animal or plant, invariably expands into every available niche. The first forests of Europe grew hundreds of millions of years ago from seedlings carried on the wind – from no one knows where. The Americas were colonised tens of thousands of years ago by primitive people who traversed a now vanished land bridge across the Bering Sea. The great nomadic nation of the Huns travelled all the way from China to the Near East, where they started a chain of events which destroyed the Roman Empire.

Returning to the world of insects, the rule may be tested by a

*The Mystery of Migration, *Dr Robin Baker (Macdonald)*

simple experiment. Leave a piece of food outdoors on a hot summer day. Within a short time, it will soon be covered with insect predators, both crawling and flying. Wherever there is a prey worth seeking, some creature will be there to claim it.

So it will be while life exists. All habitable planets in our Milky Way Galaxy of stars, provided that they support no fiercer occupant, will eventually be explored and settled by man. It is only our present deficiencies in wealth, rocket technology and computing power which prevent us from embarking immediately on a programme of interstellar colonisation.

Yet if the phenomenon of migration is universal, in the sense that any living species will tend to migrate, then alien civilisations, if they exist, have missed ample opportunities to claim the rich, lush Earth.

The story of migration on Earth, described in such detail in Dr Baker's book, reaffirms the hypothesis that the reason why no alien civilisations have visited us is that they do not exist.

Yet the migration saga has scarcely begun. Another five million years may see the whole Milky Way filled with life-forms that once originated on Earth.

Imitating Solomon

'I always like a dog', said Father Brown, 'so long as he isn't spelled backwards.'

G.K. Chesterton, *The Oracle of the Dog*

KING Solomon was supposedly able to converse with the animals, a feat which modern science has found it difficult to repeat. But several experimenters in the past few years have claimed to have had conversations with chimpanzees. Arguments about whether their claims were real or false led to furious altercations at a conference in New York.

These claims are startling. They go far beyond our everyday experience of communication with animals, in which a dog automatically responds to key words like 'heel' and 'stay', and a horse to 'Gee-up' and 'whoa!' After many years of experiment it is asserted that chimpanzees can take part in a complex conversation (complex for a chimpanzee, that is), with long strings of signals consisting of moving coloured chips on a teleprinter keyboard.

In one famous experiment at the Yerkes Regional Primate Centre in Atlanta, a group of psychologists taught two chimpanzees named Austin and Sherman to communicate from different rooms, each with a specially prepared teleprinter, and 'converse' by pushing buttons.

When, say, Austin pushed a green button, the green button would light up on Sherman's keyboard. Sherman in turn would push some button of another colour, and Austin would push another button in response. The idea was that a definite exchange of messages had taken place between the two animals and at the end of the experiment they were both rewarded with food.

If this were true, if the chimps had really been communicating by methods beyond their usual repertory of grunts and squeals, then the implication would be tremendous. We could envisage whole factory lines manned (sorry, animalled) by apes, whose wages would consist of bananas rather than money. Their only wage claims would be for an increase in the number of bananas per ape-hour, and bananas are cheap.

Alas, these claims appear likely to be untrue. They were investigated by Professor Herbert Terrace, of Columbia University, who

149

spent many futile months trying to teach 'Amelsan' (American sign language) to a chimpanzee named Nim Chimpsky. Although the first analyses of Nim's responses suggested that he was using language as humans do (roughly speaking), a closer look revealed that he was merely mimicking his teacher.

Unlike small children, with whom 'communicating' chimpanzees are often compared, Nim rarely made signs spontaneously. About 40 per cent of the time, he was just repeating the signs which Terrace made to him, without creating any sentences of his own.

His few lengthy observations contained no new information. They were merely repetitive, for example *'Give orange me give eat orange me eat orange give me eat orange give me you'*, which is a message but definitely not a sentence.

Why, then, do psychologists so strongly claim that their animals are really talking when they probably are doing nothing of the sort? Dr Thomas Sebeok, a colleague of Terrace, gave one sneering explanation at the New York conference: 'These people require success in order to obtain continued financial support for their projects, as well as personal recognition and career advancement.'

But this rude accusation sounds unconvincing. There are simply too many animal psychologists involved in the project, and it seems unlikely that so many people would jointly perpetrate such a fraud. It is much more likely that they have fallen victim to the Clever Hans syndrome.

Clever Hans was a German horse who lived at the turn of the century. His owner, Wilhelm von Osten, made the apparent discovery that Hans could do arithmetic. Von Osten would ask Hans: What are two threes?' And the horse would stamp its hoof on the ground six times.

What! A mathematical horse? People came from far and wide to watch Hans do his sums. Von Osten would ask the question. Hans would rap out the answer, and the hat would be passed round.

Then a sceptical psychologist named Oskar Pfungst noticed a curious thing. Hans only knew the correct answers so long as von Osten knew them. If von Osten had got the sum wrong, then Hans would get it wrong. There was obviously some hidden communication between man and horse, although von Osten honestly swore there was none.

Soon the answer became clear. Hans always watched his master as he answered. When he had stamped the correct number of times, von Osten involuntarily jerked his head in subconscious satisfaction. The horse saw this slight movement and ceased stamping.

The chimpanzee psychologists are likely to be making similar mistakes. They so much want their theories to be true, and the chimpanzees so much want their food-rewards, that rather than communicating, *they are going through the motions of communicating,* simply to satisfy their human teachers.

Yet there is still a remote chance that they may really have taught their animals to speak, and that Terrace and his fellow-critics may be wrong. There is an interesting postscript to the story of Clever Hans. He had a French successor named Clever Bertrand who could do the same arithmetical tricks. But unlike Clever Hans, Clever Bertrand was blind.

Rats and pigeons as economists

DO the economists, those pundits who claim to prophesy the conditions of trade, really know anything? So often are their predictions wrong that one suspects that they know very little.

Their inability to experiment makes economics a field science, like astronomy and geology, in which data comes from the real world instead of from a laboratory. But the trouble is that, unlike astronomers and geologists, scientifically speaking the economists suffer from an almost complete lack of proven principles.

That is, until 1980, when several groups of American economists did some fascinating experimental work with rats, pigeons and the inmates of a mental home. These I will describe later.

It was once thought that proven principles existed in economics that were akin to natural laws. It used to be supposed, for example, that when unemployment reached a particular level (a number that could be calculated from an equation), the inflation rate would begin to fall. But it didn't work. Although the essential connection remains undisputed, nobody has even found out what number of unemployed in a nation of given size will halt inflation. And they probably never will, for it is unlikely that any such number could be precisely determined.

Why do people have this infuriating habit of refusing to behave predictably like atoms and molecules? The reason, obviously, is that there are too few people. For instance, in Britain there are only about 20 million 'economically active' people whose behaviour one may try to predict. But there can be as many molecules of water in a teaspoon as there are teaspoonsfuls of water in the Atlantic Ocean!

No science can predict the behaviour of a single molecule, but the behaviour of millions of millions must follow the laws of probability. Yet this cannot apply to economic prediction, which concerns itself only with a few million variable units and is therefore at best a feeble science.

One group of American experimenters set out to answer the controversial question about the effect of rising wages: do they encourage people to work harder, or do they merely increase the desire for leisure? Two psychologists, Howard Bachlin, of the State University of New York, and Leonard Green, of Washington University, are trying to solve the problem by studying the behaviour of rats.

The rats in their laboratory were confronted with two buttons. One, on being pressed, released a quantity of food and the other a quantity of water. The food and the water represent 'wages', and the number of presses needed to obtain them are 'prices'.

The scientists set out to simulate a real economy by manipulating the prices and wages. They cut the amount of food that would appear on each press of the button, and they increased the quantity of water. The result? The animals *'bought'* more water and less food, a choice that humans make when commodity prices change.

152

What happens when prices fall drastically (or when wages rise, which amounts to the same thing) – in other words, if the rats could get very large amounts of food and water for very little effort? They simply lost interest in working, i.e. in pressing buttons.

In the words of the two researchers, 'non-human workers are willing to trade more income for leisure if the price is right'. The experiment, which has also been carried out with pigeons, indicates the existence of a critical level of wages beyond which higher productivity is unattainable.

Another, equally fascinating experiment involved 'token money' in a mental home on Long Island. The female inmates earned varying payments for such chores as making their own beds and an hour's work in the laundry. They were free to spend their token earnings on sweets, cigarettes, and other items.

A curious discovery was made. It turned out that the pattern of earnings of token money in a mental institution compared closely in miniature with that of income distribution in the United States as a whole. The best-paid fifth of the mental home inmates received 41.2 per cent of all earnings, compared with the national average of 41.5 per cent. The lowest-paid fifth got 7.4 per cent against 5.2 per cent in the real economy.

If experiments are possible, the analysis of small numbers of people need not necessarily produce wrong answers. Experiment may transform economics from its present state of being a set of imprecise axioms into a profound statistical science.

Perhaps more important still, people who try to study economics with computers, trying to apply pure mathematics to human behaviour, may (at least until computing is considerably more developed) be on the wrong track. The instincts and emotions of animals could prove a surer guide to the way things really happen.

The shark treatment

The shark flies the feather.

Old sea proverb

OF all dangerous animals on earth, the man-eating shark is surely the most to be feared. Sometimes weighing more than two tons and able to swim at nearly 45mph, with several rows of reserve teeth to replace those lost in its prey, it attacks without provocation and with utter recklessness.

Sharks roam the seas in their tens of millions, but only 27 of the roughly 250 different species are proven man-eaters. These range from the great white shark, the terror of the south, to the loathsome sand tiger of the American Atlantic coast, a creature that is predatory even before its birth; it sometimes eats its siblings while inside its mother's uterus.

When a man-eater is caught and cut open, the contents of its stomach often reveal its gruesome history. The stomach of one great white off the Australian coast was found to contain three overcoats, a pair of trousers, a pair of shoes, the antlers of a stag, a cow's hoof, 12 lobsters and a chicken coup. Sharks in a feeding 'frenzy' brought on by the smell of blood, will attack almost anything that moves, little caring whether it is edible.

Some of the most ruthless dictators used sharks as a means of terrorising dissenters. Ibrahim Nassir, former ruler of the Maldive Islands, kept a shark pool on a prison island, into which political suspects were lowered until they 'talked'.

Sharks can detect low-frequency sounds, far beyond the range of human hearing, that travel through the water at a speed of a mile per second, five times faster than the speed of sound in air. They can thus rush upon a wounded fish or a swimming human from a distance of hundreds of yards. Sharks stopped evolving 300 million years ago. There was no need for them to develop further. They had become perfect killing machines.

What can be done about this lurking menace? The research departments of the world's Navies have laboured for decades to produce an effective shark repellent. For at no time does success become more urgent than during a war. Tales of killings by sharks can spread demoralising fear through a whole Navy. Even the

3,000 men aboard a great aircraft carrier could be eaten if the ship sank in shark-infested waters.

Chemical stenches, metallic poisons – all the noxious products of modern laboratories – have been tried without success. For a while it seemed that no conceivable substance would stop a hungry shark.

But at last some hope has appeared. Seekers of the true shark repellent have come to realise that, to be effective, it is not enough merely to frighten off the beast; it must be paralysed or killed. For this man-made chemicals are useless. The shark's natural enemies must be turned against the shark.

Natural enemies? What sea creature would be so bold as to offer opposition to this engine of destruction? Well, none actually. It is some of the little, apparently defenceless marine creatures, both fishes and plants, that sharks fear to eat, knowing instinctively that the experience would be fatal.

By the same token, sharks will not touch seagulls or any birds that come within their reach. Hence the old sailors' saying, based on observation, *The shark flies the feather*. It is not known whether this uncharacteristic reluctance to devour is based on any good reason. But the reason is certainly known in the case of the Moses fish.

The Moses fish, a small sole found in the Gulf of Aqaba in the Red Sea – and named after the prophet who once parted that sea – exudes a milky secretion that, if fed to sharks, causes lockjaw, followed by paralysis and death as the balance of water and salts in the bloodstream is disrupted.

This secretion has been found to contain a lethal (to sharks) acidic protein called pardaxin which, whenever fed by trickery to captive sharks, induces immediate frenzy and death.

Have we therefore discovered the perfect shark repellent? Unfortunately not, because there are not enough Moses fish in the world to create an adequate pardaxin supply. Conventional chemistry cannot reproduce it artificially, for the pardaxin molecule is too large and complex. Genetic engineering may one day solve this difficulty, but in the meantime there are several other interesting and unpleasant sea creatures with which to experiment.

155

The stonefish, that sits on the ocean bed in warm seas, knowing that it need not move, for it is the most deadly of all toxic fish – this might produce an even more efficient shark poison. So also might the scorpion fish, the wasp fish, the whip fish, the helpless-looking cucumber fish, and even the coral sponge which so effectively repels parasites. Scientists are experimenting with all of them, still unsure which toxin will be both effective and available for mass production.

Although unfinished, it is an exciting story that reminds us of the climax to H.G. Wells' novel *The War of the Worlds,* in which the humblest creatures on earth, in this case the bacteria, laid low the all-powerful invaders. Some of the sea's apparently most inoffensive animals will become man's allies against the terrible predator. But they are only inoffensive because t is perilous to offend them.

A detailed report of the war against sharks may be found in the September 1981 issue of Science Digest

Don't forget the elephant

When people call this beast to mind,
They marvel more and more
At such a little tail behind
So large a trunk before.

Hilaire Belloc, *The Elephant*

IT is a sad paradox that the popularity of elephants appears to be in inverse proportion to their chances of survival. Almost anywhere in Africa, one can find tourist shops filled with waste-paper baskets made from the hooves of these slaughtered creatures, and throughout the Orient there are ivory goods on sale in quantities that imply the destruction of whole herds.

Elephants are truly the most improbable of animals, and if they didn't exist on this planet, it would be a bold science-fiction writer indeed who would dare to predict their existence on another. How unlikely a creature with a nose longer than its legs, ears wider than its behind, weighing up to 12 tons and with sufficient docility to be marshalled into a tank force that nearly overthrew the Roman legions!

If elephants are to become extinct – and it will take only a few more catastrophes like the machine-gun killing of more than 5,000 of them by Idi Amin's drunken soldiers for them to do so – then it will be well to have some lasting record of what they were like. For this, we have Mr Dan Freeman's excellent book, *Elephants: the Vanishing Giants.**

With dozens of beautiful colour illustrations, Mr Freeman delineates both modern elephants in their African and Asian forest habitats and their strange evolutionary history, from the pony-sized, long-snouted but trunkless 'moeritherium' of 70 million years ago to the gigantic, hairy Stone Age mammoth.

A vast amount of nonsense has been talked about the extinction of the mammoth. One British 'scientist' recently suggested, for example, that the discovery of dead mammoths in permafrost, with their flesh, and even their cells, intact, was evidence that they were overwhelmed by the sudden onset of an Ice Age – as if they couldn't simply migrate southwards when the climate turned cold!

The true reason for these mysterious preservations in death, as Mr Freeman points out with a splendid artist's impression, is likely to have been that the mammoths walked on layers of freezing mud, which then collapsed under their weight, burying them for thousands of years. Predators such as wild dogs would then be unable to reach their carcases.

The mammoths, so far as we know, roamed the lands bordering the Arctic, where treacherous permafrost was likely to be found, and where game was scarce, making it likely that they would be hunted by primitive man. Their extinction, therefore, is easily explained.

No such easy explanation will justify the extinction of the modern elephant, whose numbers are being inexorably reduced for profit. Nor will it quickly be forgotten that, until very recently, hundreds of the impressive and intelligent beasts were killed every year by chic Europeans for sport.

*Elephants: the Vanishing Giants, Dan Freeman (Bison, Hamlyn)

BROADCASTING

THE BBC and ITV react to the prospect of cable and satellite television, first with contemptuous disbelief that such things could ever happen, and then with petulant rage on realising their inevitability. British Telecom, meanwhile, are apparently engaged in an effort to restrict the use of private mobile radio.

Despots of the air

IN the Autumn of 1605, Ministers of James I intercepted secret letters and learned of the impending Gunpowder Plot. The king swore that never again would he or his heirs be denied the right of access to people's private correspondence. He created a postal service under the direct control of the Crown, an act of policy which is holding back much of British industry nearly four centuries later.

The Post Office, having thus started life as a spying organisation, became at the same time the monopoly which it has since remained. Monopoly has the unfortunate habit of spreading sideways, and today we have two sister-agencies, British Telecom and the Radio Regulatory Department of the Home Office, which share with the Post Office the Government monopoly in communications.

This monopoly is guarded ferociously. British Telecom is leaving the state sector, but it is struggling to retain its monopoly. We shall soon learn that a private monopoly can be just as oppressive as a public one. An antiquated telephone system is forced upon us and, according to some critics, the most shameless prevarication is employed to prevent private use of important radio frequencies.

These allegations are made in a recent study, 'Report of the Communications Crisis Committee', published by a large number of companies and commercial organisations, which shows, in horrifying detail, how unfair Government regulations and unreasonable technical specifications retard much of our technology.*

This monopoly according to the report, 'has shown itself unresponsive to the wishes of users, unaware of market considerations, inept in its assessment of the effects of developments in other countries, capable of totally disregarding international standards to whose formulation it is itself a party, and shameless in attempting to mislead Press, public and Parliament'.

One of the most unpleasant examples of this alleged chicanery is the time taken by the Home Office to issue a licence to use a particular radio frequency. Instead of taking a day or so, which one might think reasonable, the report says that it typically takes

*Report of the Communications Crisis Committee, *40 Doughty Street, London WC1.*

between four and five months.

Even by the standards of the most backward countries such conduct must be seen as not only intolerable but even suggestive that the delay is intended to discourage the applicant.

Many frequencies of the radio spectrum – the actual ether through which radio messages travel – have been allocated to the Ministry of Defence but have not been used since World War II.

Users of private mobile radio who want licences to use them are subject to the same inordinate delays.

On asking the Home Office about these difficulties I was informed that delays of more than six weeks were 'reasonable' because 'allocation of an unused frequency is a highly complex matter in which many factors have to be taken into account.'

To be as polite as I can, this explanation is simply rubbish. Any decision of a routine nature can be made in a matter of minutes if a correctly programmed computer is used, or in not more than a day if it is made solely by people.

Like all monopolies, the main reason for this one's existence is psychological. People are set in their comfortable ways and do not wish to change.

How pleasant it is to sit on innumerable interdepartmental committees and discuss these problems, to set up commissions of enquiry and do nothing while awaiting their findings – and then, when the findings appear, to set up a second commission of enquiry to explore those questions which were not covered by the first.

This, the authors of the report justly complain, is the bureaucratic way of doing things. Little will come of it, because little is intended to come of it. This active inactivity justifies the existence of administrators.

What then should be done? The report recommends the establishment of a single licensing authority to replace all these proliferating agencies, like the American Federal Communication Commission.

'Unless effective action is taken immediately,' it says, 'our future in information technology and telecommunications will be ruined, and the very business and enterprise we need will be killed off.'

But what will Auntie say to satellite TV?

THIS decade and the next are likely to see the beginnings of a vast and welcome increase in the number of television channels. There will at first be a handful of new channels, then dozens, and later perhaps hundreds, all available directly from space satellites. And the present television authorities will be able to do nothing to prevent us from receiving them in our homes.

This improvement will be achieved, very simply, by the placing of television transmitters 22,300 miles above the equator, a distance where their orbital revolutions precisely equal the Earth's own rotation. They will therefore always remain fixed over the same spot, in this 'geostationary' orbit, able to broadcast to a large area of the ground below.

The broadcasts of these satellites cannot be jammed, since the jamming device would need to operate across hundred of miles. No Air Force will be able to shoot them down, for they will not be in the Earth's air. If people want these many extra channels, and are prepared to pay for them, they will surely get them.

The most amusing part of this prospect is the conviction of the BBC and ITV authorities that it will never happen. At a recent conference of the Economics Association of Scotland, the supposed experts of these organisations reassured themselves in the manner of a horse-drawn coachman in 1900 laughing comtemptuously at the sight of a broken-down motor car.

The technical part of their argument was so absurd that a small child could see through it. Anyone wanting to receive a television broadcast from space, these experts correctly pointed out, would need to install a parabolic receiving dish some 3ft wide, pointing directly at the satellite. But, they went on to declare, the price of such a dish, with its attached electronic gadgetry, would never never drop below £300, which was more than most people can afford!

It is not merely that our generation has seen the price of pocket calculators fall from £100 to £5, and computers from more than £100,000 to £50 that makes one gasp at this assertion. It is the fact that it denies everyday observation.

With a little skill, one could even construct the necessary equipment for oneself for less than £30. Seven square feet of chicken wire can be bought for about £2 at an ironmonger's snop. Sheets of reflective aluminium to cover the chicken wire cost a few pounds more, and the receiver at the centre of the dish can be made from an old aerial. Add on a few more pounds for a frequency switching mechanism; put all these together, and you have a device that will pick up television broadcasts quite adequately from 22,300 miles. With large-scale mass production, they could sell for around £15 each.

Incredibly, these 'experts' from British television seemed unaware of Arthur C. Clarke's experiments in India in the 1970s when he and his colleagues toured some 5,000 villages, teaching people to make receiving dishes from chicken wire and use them to watch television broadcasts from space. But ordinary people knew of Clarke's Indian work, for it had been reported on British television.

Now for the cultural side of this business. Mr Brian Wenham, chairman of a BBC working party on cable and satellite broadcasting – who must be assumed to have considered the question – solemnly assured the conference that television audiences had indicated 'over the years' that they preferred home-produced material to that imported from abroad. Therefore, he inferred, they would not wish to watch satellite broadcasts.

Mr Wenham's logic is hard to follow. Why does he assume that satellite broadcasts would consist of material 'imported from abroad'? I cannot think of any reason why they should. Only a lunatic would think it a good idea to beam down into British homes a constant flow of European programmes which very few people would watch.

But it will not happen like this. Companies wanting to reach the British television market, whether they are foreign-owned or not, will employ British people, who know their home market, to compose their programmes. They will pay British advertising agencies to make their commercials, and they will tend only to advertise goods that are available in British shops. Mr Wenham and his friends seem to think that everything from outer space

163

must be 'foreign'. But it does not have to be. People will care more about the cultural origin of the programme than its means of transmission.

What of the legal position? Surely, it will be argued, the foreign satellite-owners would not dare to broadcast into our homes without the agreement of the British Government? And would not the BBC and ITV make sure that the Government made no such agreement unless the contents of the broadcasts were under their own control?

Any such moves are likely to be futile. For Britain will be in a weak bargaining position. With no capability to launch space satellites of our own, we rely on the facilities of others. It is true that we could threaten to withdraw our 2.5 per cent investment in the European Ariane rocket, but the shortfall would be quickly made up. In short, there is nothing to prevent sufficiently wealthy organisations from broadcasting uncensored.

In North America, this has already happened. Despite the futile objections of the established networks, there are now 12 geostationary satellites over the United States and Canada, carrying a total of 68 television channels, a number which is expected to increase rapidly.

We cannot tune in to these North American satellites, for they are far over our western horizon. But after the mid 1980s several European broadcasting satellites are likely to have been launched.

What will their programmes mainly consist of? One cannot predict their contents precisely, but it seems that the most promising possibility is of high quality films, made specially for television, which might be made simultaneously in several languages. At the very least, it will make a change from compulsory Party Political Broadcasts.

MISCELLANEOUS

Roman emperors poisoned by lead in their wine; underground trains that could cruise at 6,000mph; why too much attention to public safety can endanger the public; how to unmask anonymous writers; the treasures of Antarctica; an easy way to set up in business as a weather-forecaster; the eternal muddle over metric and imperial systems of measurement; and, my favourite chapter of this book, a glimpse of the unscrupulous founder of modern science, that shady politician, Francis Bacon.

Lead-crazed Romans

. . . a slow and secret poison in the vitals of the empire

Edward Gibbon

IT has long been known that the aristocrats of imperial Rome were in the habit of murdering each other by poison. But scholars have only recently discovered the extent to which they poisoned themselves accidently, by imbibing lead at a level far above today's safety limits.

Excessive lead poisoning among the most rich and powerful Romans, it is now agreed, may have contributed to the downfall of the Roman Empire. In a fascinating article in the March 1983 *New England Journal of Medicine,* Dr Kerome O. Nriagu of Canada's National Water Institute, surveys the evidence and reaches some trenchant conclusions.

The Romans drank prodigious amounts of wine, and what they drank was freely mixed with grape syrup which had been boiled and then left to simmer in lead pots or lead-lined kettles. The grape syrup according to Columella, an epicure of the first century AD, was needed for the colour, sweetness and preservation of the wine. And, his deadly prescription went on, the simmering down must be done in a copper vessel lined with lead, for those lined with brass or bronze throw off a copper rust which has a disagreeable flavour.

Preparation of this delicious but foul syrup, says Dr Nriagu, would have produced lead concentrations of between 240 and 1,000 milligrams per litre. One teaspoon of such a mixture, he believes, would have been more than enough to cause chronic lead poisoning.

A frequent result of lead poisoning is gout, produced by malfunctioning of the kidneys, a painful condition which the Romans shared with many British aristocrats of the eighteenth century, who also fortified their wines in the style recommended by Columella.

Gout, by the pain it brings, can cause outbreaks of irascible behaviour in its victim. But the British aristocrats of 200 years ago were harmless creatures compared with the overlords of Rome. They had no domestic slaves to flog or slaughter. Except in rare official capacities, they had no power to sentence anyone to death. They no doubt had moods of savage irrationality, but they could not ease their feelings by setting fire to cities or leading armies against the State.

Viewed in this light, the incessant outbreaks of internecine bloodlust that characterised the Roman Empire may have been induced by gout and by that deterioration of the brain that lead poisoning can also bring.

There are too many examples of pointless butcheries in imperial history for this possibility to be set aside. The generals Marius and Sulla, fighting a civil war for no apparent reason other than mutual hatred and bad temper, in turn captured the city; and each with no political motive, slew every citizen their soldiers could lay hands on. Mark Antony, a phenomenal drinker, insisted on murdering half the Senate after Julius Caesar's death, even though he knew most of them were innocent of it.

The emperor Tiberius was a special case. He drank in the way he killed – with loving selection. According to one lampoon of the time, he moderated his drinking in order to concentrate on executions:

He is no drunkard now,
As he was drunkard then.
He fills him up with a richer cup,
The blood of murdered men.

So it went on through the long, dark centuries, as emperor and pro-consuls drenched their kidneys and brains with lead and the rest of the world with blood.

It is the lack of motive behind these crimes that so irresistibly suggests brain poisoning. How else can one explain the fury of Caligula, who wished the Roman people had but one neck so that he could slice it through? The same may be asked of the incendiary violence of Nero; the calculated sadism of Domitian; the ungovernable temper of Septimius Severus who had dead Senators dug up from their graves and dangled outside their homes; of Commodus, and many like him, who needed the spectacle of suffering as other people need air. Of mad Elagabalus, who drank so voraciously that people said he drank from a

167

swimming pool. And of the brutal Maximin, who entertained a specially murderous hatred for the friends of his youth, because they had committed the crime of witnessing his youth.

Research by Dr Nriagu shows that about 19 out of the 30 emperors between Augustus and Elagabalus, roughly two-thirds, had a taste for lead-tainted wine. Skeletons recently found in a Roman cemetery in Cirencester, with a lead concentration 10 times what is normal today, prove that lesser citizens copied them. Some scholars, such as Pliny the Elder, had warned that the mixture was dangerous, but the warnings were ignored. Imperial Rome drank its way into the Dark Ages.

Academic historians too often ignore the medical conditions of their subjects but Dr Nriagu appears to have filled an important gap in historical knowledge. One hopes that others will learn to do the same, and instead of confining themselves to the minutiae of treaties, will ask instead how much a key official had to drink that day.

The age of the super-train

Don't be too optimistic, that light at the end of the tunnel may be an oncoming train.

Robert Salter

HOW would most people react to the prospect of travelling from London to New York by train ten times faster than a jumbo jet, and arriving within 58 minutes for a fare of about £1? In much the same way, no doubt, as the Victorians would have reacted to the notion that their great grand-children would one day fly the Atlantic at 580mph.

Underground trains with a cruising speed of at least 6,000mph are the dream of an ex-missile engineer Robert M. Salter, now at the Rand Corporation in California.

Such a network of super-trains, passing alike under land and sea, would remove most of the noise and pollution from our skies. No longer would thousands of square miles of land have to be given up for international airports. Perhaps most important of all, it would save countless billions of pounds worth of petroleum fuel.

A jumbo jet, for example, burns more than 3,000 gallons of fuel merely to reach its cruising altitude of 33,000 feet, whereas Dr Salter believes that his 'Planetran' locomotives would recycle about 97 per cent of their energy and travel half way across the world for the price of a few pounds.

A bizarre claim, it might seem. How would it work? Quite simply, by magnetism. The evolution of railway trains has probably come to an end. Any speed substantially greater than about 200mph would put pressure on the rails beyond the limits of tolerance. The next step forward is for trains that move without rails, that travel, in other words, on electromagnetic fields.

The missiles which Dr Salter was designing back in 1957 would attain near orbital speed as they flew through the vacuum of space from one continent to another. Why not construct underground train tunnels, he mused, that would simulate the conditions of space by being almost an airless vacuum? Without the friction of air, the only speed limit would be the curvature of the Earth.

Even in the open air, a train can attain extremely high speeds by riding on magnetic levitation, a technique known as 'maglev'. In 1980 in Japan, an experimental maglev train reached 325 mph as it glided a foot above the track.

The technique is simple. As most people know, like-poles in a bar magnet repel one another. In the Japanese experiment, as in Dr Salter's planned Planetran project, there are like-poles both in magnets in the train and in magnets in the track beneath. The repulsive force between them is enough to lift the train.

In practice, the Planetran system would be somewhat more complicated. Instead of bar magnets, the trains would be elevated by the field of electromagnetic coils that would be cooled with liquid helium to minus 453°F., just above absolute zero.

In this way, the coils would become super-conductors. All resistance to the flow of electricity would disappear, and a magnetic field would be produced that propelled the train even after the source of power had been cut off.

169

This magnetic field would move faster and faster down the vacuum tunnel, with the train riding on it, much as a surf boarder rides a wave.

It must be admitted that doing this on Earth may be prohibitively expensive. To construct a supersonic underground train service that linked all the major cities of the world, digging deep beneath the bed of the oceans, could involve a capital construction cost of more than £100,000 million. Even spread over half a century, this would be a significant fraction of the gross world product; and because the system, once built, would be so cheap to operate, there might be difficulties in raising money to make it profitable.

Moreover, the imminent prospect of the introduction of water-based liquid hydrogen fuel for jet aircraft removes much of the economic attraction for Planetran. Yet there are two ways in which it may come about.

One is to do it in a relatively local area, such as Europe or the continental United States, where an enormous number of deep tunnels exist that could be modified to accomodate Planetran, and where the profusion of large airports is an almost intolerable nuisance.

The second way is even more probable. It will be built within the next hundred years, when large numbers of people are likely to be living on the moon and perhaps on Mars. Aircraft cannot fly from place to place on these worlds – for there is little or no air for them to fly on. For local expeditions, a Salter network of supersonic trains would be the most efficient means of extra-terrestrial transport.

A head-on collision between two trains moving at 6,000mph or more would be truly catastrophic. The kinetic energy unleashed by the impact could make a surface crater many miles wide as it tore open a gigantic hole in the ground. If it happened beneath a city, the effect would be like the fall of a 200-ton asteroid.

And so I am happy to accept Dr Salter's assurance that so wide an array of safety devices will be built that such a thing will be impossible. Nevertheless, some people will always worry about unidentified lights at the end of the tunnel.

When safety means danger

And you all know, security
Is mortals' chiefest enemy.

Macbeth

FOR many decades, a large number of legislators have been demanding that consumer products, from drugs to aircraft, should be guaranteed to be '100 per cent safe' before being used – and uncounted tens of thousands of people have died because of the insistence on this principle.

It is a principle only now beginning to be understood, that too much attention to public safety tends to create public danger. There have been several instances of this in the last few years.

One of the most scandalous was the American Government's decision in 1979 to ground all DC-10 aircraft for several months after a crash in Chicago caused by a fault in the tail of one of these planes, in order to enable all other American DC-10s to be investigated. During this long period, DC-10s were forbidden to fly in America or land there after flights from abroad.

A sensible decision, one might think. Why should I call it 'scandalous'? The reason is plain. Huge numbers of air travellers were forced to travel by much more dangerous conveyances.

In the words of a distinguished expert on risk and safety, Professor Sir Ernest Titterton, of the Australian National University, 'People who would have travelled in America by those aircraft now went by road, which was 100 times more dangerous, or by rail, 2.5 times more dangerous, or on older, smaller, much less-safe aircraft.'

In addition, the grounding of the DC-10s which, statistically, were so safe that they only killed one person per 400 million passenger miles, inflicted catastrophic financial damage on the world's airline industries, as well as contributing to the bankruptcy of Sir Freddie Laker. Perhaps significantly, other governments refused to back the decision – to the astonishment and fury of officials of the American Aviation Administration, who consider themselves the supreme authorities on air safety.

Still worse, because it arose from even greater official stupidity, was the American decision to shut down all nuclear power

171

stations with the same design as the reactor at Three Mile Island, Pennsylvania, which in 1980 suffered a small leak of radiation which injured nobody.

This decision, taken in response to what Sir Ernest rightly calls 'irrational and exaggerated media reporting' of the Three Mile Island incident, produced in his opinion, an 'incalculable number of deaths'. For it created instantly a much greater dependence on coal-fired electricity, which is a vastly greater threat to public health than nuclear power.

'The closure of these nuclear power stations,' he explained, 'will therefore lead to some 100 deaths per year, some 125,000 cases per year of chronic respiratory disease, a million person days of aggravated heart-lung symptoms, and about £12 million worth of assorted property damage.'

One of the daftest notions ever put forward, received with solemn seriousness by many government agencies, was the Great Aerosol Scare of 1974. Two scientists at the University of California proposed the theory that the contents of aerosol spray cans, known chemically as chlorofluorocarbons, could on rising into the stratosphere seriously damage the Earth's layer of ozone, or triatomic oxygen, which protects us from dangerous ultraviolet radiation from the Sun.

Chemically, the theory made sense. Compounds in aerosols do indeed break down ozone. But statistically, the idea was the most far-fetched rubbish. The ozone layer is spread over an area of tens of millions of square miles. The quantity of aerosol compounds being released daily by the human race is so small that it would take thousands of years for them to have any noticeable effect.

But what happened? We were treated to the same routine as with the DC-10s and the aftermath of Three Mile Island. The United States announced a ban on aerosol products with which other governments refused to comply. They mistrusted the American mania for regulation. In the words of Professor Sir Richard Scorer, then Chairman of Britain's Clean Air Council, 'these civil servants have been given power to regulate, and regulate they will. They don't listen to scientific arguments. They have already made their minds up. But the truth is that the Earth's atmosphere is a very complicated system and the more complicated a system is the more robust it usually is.'

This is no need here to tell in detail the story of the American ban on saccharins, for it follows the same lines as those of the DC-10s, the power stations and the aerosols. A gigantic intake of saccharines, which no human would dream of consuming, induced cancer in a rat, and that was enough for the American regulators. Out went the non-fattening saccharines, and people turned massively to sugar, with all the predictable consequences in obesity, heart disease, rotting teeth and the risk to the lives of diabetics.

British governments can be just as prone to reckless safety regulation, though on a less-spectacular scale than the Americans. According to a report published recently by the Royal Society, we have the same tendency to insure against improbable disasters and ignore the more likely ones – a good example was the new building regulations introduced after the Ronan Point explosion (which killed several council tenants), which cost about £20 million for each life lost.

The Royal Society team, led by Sir Frederick Warner, is making one sensible proposal that is bound to be hotly resisted: that laboratory tests for new drugs should be made less stringent, so, as Sir Frederick explains, 'a new medicine can come quickly into public use without waiting years while people die for the lack of it.'

What of the future? The director of the Office of Health Economics, Professor George Teeling-Smith, predicts that within the next 40 years we shall have learned to control many of the diseases that now threaten us, and among these he lists most cancers, virus infections ranging from influenza to shingles, multiple sclerosis, rheumatoid arthritis and senile dementia.

But this optimistic dream, he points out, will come true if, and only if, we restrain those legislators who insist that drugs must be '100 per cent safe'. Whatever that may mean.

173

Weather forecasting on the cheap

Oh, what a blamed uncertain thing
  This pesky weather is!
It blew and snew and then it thew
  And now, by jing, it's friz!

Philander Johnson, *Shooting Stars*

ONE of the hottest summers on record again arouses interest in the weather. Why is it now so unpleasant and at other times so agreeable? An excellent book by Mr Gunter Roth does much to explain the cause and effect of all possible weather conditions – and helps us also to understand the technical jargon of the nightly forecasts.*

It is a book produced in an original style that is aptly suited to its subject. Instead of the usual essay of 80,000 words, in which information is scattered around in different chapters, Mr Roth has written an illustrated encyclopedia of only 256 pages, with a fine colour photograph to illustrate each state of the weather, showing its appearance, its cause and its effect.

It is full of interesting little details that many people do not know. The colder the air temperature, for example, the more likely it is to rain. At 68°F., a cubic yard of air contains 0.61 ounces of water vapour. But as it gets colder, the water vapour condenses into rain droplets. When the temperature has dropped to freezing point, the quantity of water vapour in each cubic yard will be only 0.17 ounces, for the rest is likely to have fallen as rain.

Certain cause-and-effect relationships may be simple, but to forecast weather accurately is immensely complex. One problem is that no computer is as yet fast enough to process all the huge quantities of information in the time available for the forecast. What use is a forecast of the weather five days ahead if it takes ten days to make the forecast? Another problem, in Britain at least, is the loss of forecasting staff due to budget cuts.

Since December 1980, we no longer have had long-range weather forecasts, for the staff at the Meteorological Office at

*Collins Guide to the Weather, *Gunther Roth (Collins)*

Bracknell engaged in this work has been cut from 30 to 23, a number considered too small for the job. Weather forecasters are greatly relieved at the change, since the long-range forecasts however accurate they might have been, seemed to have done little but infuriate the public.

'You incompetent idiots! You may have told us that the month would be half a degree colder than average (and a fat lot of use that information is), but you never told us it would pour with rain throughout Bank Holiday.'

This sort of reaction to the long-range forecasts was getting down the morale of the Met. Office staff. Joe Bloggs, setting out on his camping tour, could not understand that information about next month's average temperature, although useless to him, was vitally important to such organisations as the British Gas Corporation, to whom the forecasts are still confidentially circulated.

Imagine poor Joe Bloggs and his family, growing increasingly bad-tempered as they sit shivering in a tent with rain pouring in through the cracks on a day they had been told would be sunny.

Why couldn't they have been warned?

The reason is simple: Certain facts about future weather are more difficult to forecast than others. Generally speaking, they go in a descending order, from the easiest to predict to the hardest:

1. The pattern of air pressure.
2. The direction and strength of winds.
3. The general level of temperature.
4. The overall level of rainfall.
5. Surface temperature.
6. The variability of clouds.

The predictions which most affect the well-being of Joe Bloggs are numbers 5 and 6, the most difficult to forecast of all! For they depend on each other, and they alone can decide whether there will be sunshine or rain in a particular region. Since Joe cannot camp in more than one place at a time, the success of his holiday can depend on the most minute changes in local conditions which no mind, whether human or electronic, can yet predict.

There are many superstitions about the weather. One should not take too seriously the belief that a red sky at night implies fine weather the next day – or that shepherds should take alarm at the sight of a red sky in the morning.

A more interesting legend is that of St Swithin's day, 15 July. It has long been believed that the weather on that day will continue for the next 40 days – which happened at Winchester in 964 AD, the year of the saint's burial, when it allegedly rained without cease from 15 July to 25 August. But the myth of St Swithin's day is exposed by the change to the Gregorian Calendar in 1752, when our days were moved forward by 13. The weather, after all, cannot be expected to know our calendar.

Can one make money from setting up a private weather forecasting agency? People with principles might find it hard to keep up with the sophistication of the Met. Office, but very little knowledge is required to give reasonably convincing forecasts with a better than even chance of being right.

An unscrupulous weather forecaster would need only to predict a continuation of today's weather into tomorrow, and more than half the time he would be right. He could not equal the 85 per cent accuracy claimed by the Met. Office, but a 60 per cent success rate, if bolstered by some slick advertising, and confined to a local area, would satisfy some innocent subscribers.

For he would only be taking illicit advantage of the fact that certain types of weather have a tendency to continue for many days.

Graffiti in the desert

ALONG the southern coast of Peru lies the vast and arid Nazca Desert. Across its black gravelly surface are huge drawings in the sand, of frogs, birds, monkeys, of triangles, rectangles and long straight lines, all made by some long forgotten race.

Who were these people? Why and how did they decorate the landscape in this extraordinary fashion? The survival of these drawings after thousands of years in this darkly desolate place has turned these questions into a profound mystery.

The Nazca drawings became well known in 1970 when the ex-innkeeper Erich von Daniken, in his bestselling but ridiculous book, *Chariots of the Gods,* alleged they were the work of creatures from outer space. One of the diagrams had the general shape of an airport parking area; and this to von Daniken, was proof that it was an ancient airport – or spaceport.

But it is unlikely to have been anything of the kind. A heavy vehicle would instantly have become stuck in the soft sand, and had aliens with interstellar technology ever visited the earth, they would certainly have left more obvious traces of their visit.

The drawings are nevertheless highly spectacular, many of them hundreds of feet across. The paradox is that they seem to have been drawn as if their artists intended them to be seen only from the air – while nobody at the time could fly.

They are an extreme version of certain types of gigantic 'works of art' found in Britain, also of unknown age. One thinks of the huge white horse on the Berkshire Downs near Uffington, allegedly made by the Anglo-Saxon leader Hengist (whose name means 'stallion'), and the blatantly sexual giant on a hillside above Cerne Abbas in Dorset.

On hillsides made of chalk, with a thin covering of turf, it must have been easy to remove the turf and leave the desired shape in the form of exposed chalk, perhaps a week's work by a score or so people with simple tools.

These hillsides in Britain are steep, and one can easily make out the shape of the drawings from the next door field. But the Nazca Desert is largely flat, and it is difficult to see the depicted shapes except from the air. How and for whom were they drawn? This is the riddle that has attracted the cynical journalism of von Daniken and baffled mathematicians, archaeologists and astronomers ever since pilots discovered them.

Astronomers in particular have made tremendous efforts to find at Nazca some celestial significance. The American Gerald Hawkins, who has used a computer to solve the mysteries of Stonehenge, tried to do the same in Peru. Success utterly eluded him. Did the Nazca lines point to the rising or setting of the Sun or moon? Only in 39 out of 186 cases did they do so, which statistically was insignificant.

177

Hawkins went further, eager to find some connection between the drawings and the sky. He turned to the stars. Uncertain when the Nazca people had lived, he fed into his computer catalogues of star positions dating back to 10,000 BC. Again, he learned nothing. The Nazca alignments with the stars were too few at any period in history to justify the suspicion that they were intended

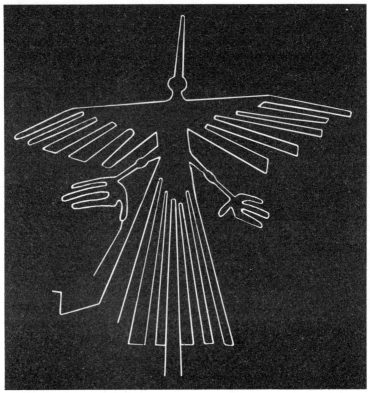

Man-made art: to prove the absurdity of claims that the famous Peruvian desert drawings could only have been made by creatures from outer space, this giant bird was reproduced in a Kentucky field with no tools more complex than human muscles and a few balls of string. (Courtesy of The Sceptical Enquirer)

as an astronomical calendar.

To the more direct mind of Arthur C. Clarke, the problem is somewhat exaggerated. Many primitive peoples have imagined they saw pictures among the stars, so why should they not have tried to return the compliment and transmit their own pictures to the stars? The mystery, he says, may be no more complicated than man's age-old desire to leave some record of his existence, saying KILROY WAS HERE, on a scale that fills the landscape.

Few ideas seem sillier than those of von Daniken and other fantasy-mongers that ancient peoples could not have built magnificent structures and soaring works of art unaided by supernatural power or advanced technology. Roman aquaducts and French cathedrals were built without modern engineering tools, so why should the Nazca drawings be thought to require the assistance of little green men from Alpha Centauri?

As James Randi, the American magician and scholar, remarks in his magnificent book *Flim-Flam**, 'Try as he may, von Daniken cannot diminish the works created by greater men than he. For every giant, there is a little man to kick at his ankles. But the great accomplishments of long ago remain.'

The mystery of how the Nazca drawings were made appears to have been solved by Joe Nickell, of the University of Kentucky. Following the methods suggested by the British explorer Tony Morrison, he used simple geometry, marking out lines with lengths of twine and removed the necessary gravel, to reproduce in a Kentucky field the 440-foot figure of a giant bird, copied to scale from the same bird at Nacza and shown on the facing page. With several helpers, it was the work of a few days.

---

*Flim-Flam: E.S.P., Unicorns and Other Delusions, *James Randi (Prometheus Books; available from Michael Hutchinson, 10 Crescent View, Loughton, Essex)*

Now you see it, now you don't

So she went into the garden to cut a cabbage leaf to make an apple pie; and at the same time, a great she-bear coming up the street pops its head into the shop – 'what! No soap?' So he died, and she very imprudently married the barber: and there were present the Picninnies, and the Joblilies, and the Garyulies, and the great Panjandrum himself, with the little round button at top. And they all fell to playing the game of 'catch as catch can', till the gunpowder ran out at the heels of their boots.

Samuel Foote

THIS gem of nonsense prose by Footé, the great eighteenth-century dramatist, was written to challenge a friend who had boasted that he could remember any passage by heart.

This passage is difficult to remember for one particular reason: it is full of non-sequiturs: it makes no concession to causality, the principle that an event cannot occur without a preceding cause.

Why, for example did the shopkeeper die? Was it from mortification at having no soap to sell to the she-bear? Did the animal kill him in disappointed rage? Or did he die of shock at being expected to eat an apple pie that was made of cabbage leaves? And was it the great 'imprudence' of the subsequent marriage that persuaded the wedding guests to conceal gunpowder in their boots? And who married the barber anyway, the housewife or the she-bear? These questions are deliberately left unanswered. The lack of causality in the story leaves us with the same sense of weirdness that we would get from reading:

1. London Bridge has fallen down and normal service will be resumed shortly.

2. The aircraft landed safely, and all the passengers were killed.

The unnerving effects of these literary jokes will give some idea of how it must feel to be confronted by an apparently real breakdown of causality. This is now the situation in the branch of physics known as quantum mechanics. Things have happened which ought not to happen. There have been events without a cause. This does not mean merely that their cause is still undiscovered: there may even have been no cause.

Before proceeding immediately into this jungle of impossible things, let us take a quick look at the Uncertainty Principle, discovered by Werner Heisenberg in 1927, which lies at the root of quantum mechanics. This principle states that it is impossible to obtain complete knowledge about both the position and the speed of an electron, the particle that spins around the nucleus of an atom. The more one learns about one, the less one can know about the other. Again, this does not mean that our measuring instruments are inadequate. It means that the complete information does not exist.

Knowledge is erased by the presence of the observer. To discover the exact movements of the electron one needs, so to speak, to see it. The act of seeing it requires light, in the form of photon particles. The mass of the photons actually moves the electron.

Suppose you have left a football out of doors at night. You go out and look for it with a torch. Now, to an infinitesimal degree, that football is not in the same place when it is in darkness as it is when your torch shines on it. The torchlight itself moves the football. The movement is of course far too slight to be detected by any instrument but it is not zero. In darkness therefore, the absolutely exact whereabouts of the football are an impenetrable mystery.

So it is with electrons which, being so much smaller and lighter than footballs, are much more easily moved around by the photons that are needed to observe them. The behaviour of the 'microworld' of electrons is thus quite different from that of the 'macroworld' of cathedrals, railway stations and office desks. Large objects stay where they are supposed to stay, but with tiny objects, the structure of reality is continuously and unpredictably changing.

The strangest of all phenomena of quantum mechanics is the Einstein–Podolsky–Rosen paradox, predicted by those three savants in 1935.

The paradox is this: if two sub-atomic particles that have previously been in contact are later separated by a great distance, then a measurement carried out on one of them will bring an instantaneous change of state on the other, even though it may be

181

thousands of miles away! No message has passed between the two particles. The thing simply happens. Einstein himself was profoundly disturbed by this inexplicable action-at-a-distance, which he called 'spooky', because it violates his special theory of relativity which states that no message can travel faster than light.

What is the explanation? How does one particle 'know' instantly what is happening to its distant twin? Nobody knows; one can only speculate. According to one hypothesis, nothing is absolutely real, and the universe exists only because we are present to observe it. In the words of Ronald Knox:

> *There once was a man who said 'God*
> *Must think it exceedingly odd*
> *If he finds that this tree*
> *Continues to be*
> *When there's no one about in the Quad.'*

Which provoked the anonymous reply:

> *Dear Sir, your astonishment's odd.*
> *I am always about in the Quad.*
> *And that's why the tree*
> *Will continue to be*
> *Since observed by yours faithfully, God.*

The truth must be somewhere in between these two extreme positions. But to ask exactly where is like asking which of Foote's characters married the barber.

The treasures of Antarctica

Lands doomed by nature to perpetual frigidness, never to feel the warmth of the sun's rays, whose horrible and savage aspect I have not words to describe.

Captain Cook in 1774

THE vast and desolate continent of Antarctica, here described by one of its earliest visitors, is beginning to lure the attention of the great powers, who increasingly see in it a treasury of wealth.

Fishing stocks to feed the hungry in their tens of millions, untold resources of oil and natural gas, spectacular landscapes for the tourist, a commanding strategic position in the event of war; all these aspects are attracting people to a terrain that is as strange as an alien planet.

What is it like, this extraordinary place? Imagine five and a half million square miles of ice-covered plateau and partly unexplored mountain ranges, nearly twice as large as Australia, with not a single tree or bush or trace of greenery and swept by raging blizzards which can reduce the temperatures to minus 60° F. more than 90 degrees of frost.

Antarctica is one of the geographic marvels of the world. It is as if some giant had created it by pressing his thumb into the North Polar regions to produce a great dent in the globe, with a compensating bulge in the far south.

For the antipodes of the northern and southern ice caps are curiously opposite. They are nearly equal in size. Yet the north is a vast ocean with an average depth of 4,200 feet, while the south is a plateau with a mean elevation of 6,000 feet – as though the one was composed of land mass stolen from the other. The greatest depth of the Arctic Ocean, some 17,500 feet, has its opposite in Antarctica's highest mountain, Vinsom Massif, which rises 16,800 feet above the sea. Even the very shapes of the two polar icecaps have a strange similarity, so that a map of one could be imposed on the other with little overlapping.

It was not always like this. Sixty million years ago, a mere hundredth of the timescale of the Earth's history, Antarctica was a tropical land, part of a huge primaeval continent comprising what is now Africa, South America, Australia and India, that scientists call Gondwanaland. Then the sea floor began to move, wrenching Gondwanaland into fragments. Today, the fossils of tropical plants and animals are sometimes found beneath the ice, showing that this frozen wilderness once enjoyed a balmier climate.

The very depth of the ice, in some places two miles thick, makes it difficult to imagine the extraction of mineral and petroleum resources from the rocky Antarctica crust. For the ice is constantly moving. A drilling hole might cease to exist 24 hours after it was bored. But the situation in the waters surrounding Antarctica may be more promising. After the voyage of the exploration ship Glomar Challenger in 1973, it was estimated that hundreds of millions of tons of oil and natural gas may lie there undiscovered.

The day must surely come when it will be practical and economic to drill for it. But this will not be easy. Oil rigs, even those built to withstand winter storms in the North Sea, would be in daily danger of being crushed by wandering icebergs whose

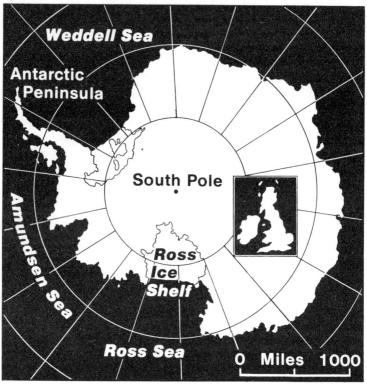

The Antarctic, with the British Isles shown to the same scale

184

size can approach 100 square miles. A better prospect of success lies in oil-drilling ships with a strength and manoeuvrability few vessels yet possess.

It may seem an odd proposition that Antarctica could one day feed the Third World; but the idea has been seriously put forward. The hopeful new food source is krill, match-sized shrimps, at present the diet of the great blue whale, the largest animal on earth, which gets through three tons of krill every day.

Krill is the most plentiful source of protein to be found in the oceans. But attempts to sell it on a large scale for human consumption have so far been unsuccessful. It nourishes, but it lacks taste. Culinary imagination is still needed to make it palatable.

But it is in the strategic sense that Antarctica looms in importance. The Antarctic Treaty of 1961 forbids any military use of the continent, but treaties are often ignored in time of war. Imagine the Panama Canal blocked and the tip of South America in the hands of a hostile power. Only those nations with Antarctic bases could then send ships between the Altantic and the Pacific.

Many hundreds of visitors are likely to go southwards in the next decade in response to this new awakening of interest. The spectacular television pictures of South Georgia that appeared during the Falklands War may have had something to do with this. But for whatever reason, the highest, coldest, stormiest, driest and least known continent on Earth has suddenly come to seem the most exciting of frontiers.

Antarctica is in many ways a more dangerous place to live than on the moon, since on the moon there are no storms or unpredictable conditions. But the South Polar regions are considerably more beautiful and more accessible. Nowhere else on this planet can one find ice-cliffs hundreds of feet wide and hundreds of miles long as flat as the architecture of a city. Nowhere else can one find such flat snowfields, stretching out distance upon distance for horizon after horizon, beneath the melancholy, red glow of the midnight sun.

Little words to unmask big authors

THE historian Josephus, writing his book on the Jewish-Roman war of the 1st century AD, realised suddenly to his horror that he had not the faintest idea how to describe a battle. And so he copied his battle scenes from Thucydides, who had been recounting another war 500 years before, changing only the proper names. Surely no one would find out – and no one did, until the invention of the computer.

Until recently people who wanted to conceal their authorship of documents did so by the simple expedient of omitting their signatures or adding somebody else's. It seemed so easy. So long as you imitated the outward manifestations of another person's style, who was going to tell that it was not genuine?

Shakespeare thought himself as safe as Josephus did when he borrowed the last act of *Henry VI Part I* from one of his friends. The early Christian fathers thought it a pity that St Paul had died before writing an admonitory letter to the Hebrews. But that was no problem. They just wrote one and added his name to it. The KGB feared no detection when they inserted in a book by Kim Philby several chapters of their own, which were designed to cause dissension between Britain and America.

But most of these frauds have now been found out; Shakespeare's by Mr Thomas Merriam, a Basingstoke lecturer, using a computer that cost him less than £100; the true and faked letters of St Paul have been identified by Dr Andrew Morton, a Scottish minister; and the KGB's misquotations of Philby by the CIA, using the same method of detection as Mr Merriam.

How is it done? Not by looking at the average length of sentences or the choice and profusion of adjectives and adverbs, for these things can be changed according to the subject matter. It is the incidence of the little phrases, such as 'and the', 'but the', 'from the', 'of the', and the frequency of sentences beginning with prepositions like 'But', 'It', and so forth, which are seldom or never changed. Their pattern is a thing ingrained in a writer by subconscious habit, and alters little from early adulthood to dotage.

Consider an inflammatory political pamphlet filled with stirring calls for action. On another occasion its author writes a begging letter to his bank manager. The wording of the two seems wholly different; for one demands the wrecking of Cabinets while the other complains of the tardiness of debtors. Neither of these writings appears to have anything in common with the other.

But to the cold inspection of 'stylometrics' the common authorship will be plain to see. To illustrate this, I have made out rough stylometric charts of two well-known British writers, Mr Julian Critchley MP and Mr Auberon Waugh.

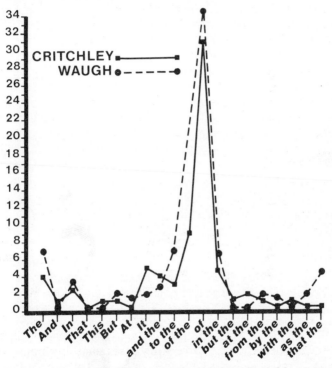

A stylometric chart of the writing of Julian Critchley MP and Auberon Waugh: the most obvious differences between Critchley and Waugh are in Waugh's repeated use of 'that the' and Critchley's habit of starting sentences with 'It'. The plotting of more words and phrases would produce still more accurate graphs.

Messrs Critchley and Waugh made excellent subjects for my experiment, for they are both 'general purpose writers'; they turn their pens to any subject that interests them. Mr Waugh, for instance, writes scholarly articles in the Spectator, while the tone of his column in the Sunday Telegraph is light and irreverent. Consider these two passages from Mr Waugh:

1. *The fact that those utopian movements which survive eventually re-write their histories to pose as the great champions of the family illustrates better than anything else how the family will always win against such aberrant fantasies.*

2. *Sooner or later, someone in Paris is going to pick up the civil administration of the Cote d'Azur and shake it until all the fleas drop out.*

Had the articles been unsigned, would it not take a magician to detect the common authorship? On the strength of these short samples, indeed it would. But when one analyses the entire texts, each of which is nearly 2,000 words, and counts the incidence of eight prepositions and 12 prepositional phrases, one reaches the conclusion that the expert on utopian movements in the first journal and the expert on shady doings on the Cote d'Azur is likely to be the same person.

Two years ago, Mr Critchley wrote a lengthy anonymous newspaper article criticising Mrs Thatcher's economic policies. Angry Government whips had no success in unmasking him until apparently, somebody 'sneaked'. How much easier would the whips have found their task of detection if, instead of saying: 'Did you write it?' to nearly every Tory MP they met, they had used the surer method of stylometrics! They needed only to draw up a short list of suspects, obtain a few of their published writings which exceeded 1,000 words, and instruct a computer to count the little words and phrases.

This I have done as demonstrated in the graph. I took two articles by Mr Waugh, and three by Mr Critchley, including his anonymous piece, and even in this rough-and-ready experiment, the individuality of each writer is as plain as if it had been emblazoned in neon lights.

Doing this *without a computer* is no fun. To go through people's prose, often using a smudgy photocopy, ignoring the subject matter, and counting all the 'in the's' and 'for the's' and the rest, marking each with a different underlining sign to avoid subsequent confusion, is a labour so exhausting, so time-consuming and essentially so boring that the keenest literary detective must quail from it. But the computer does it in seconds.

Yet there are pitfalls for the inexperienced. How many times for example, does the word 'and' occur in the phrase: 'Handsome Andrew and Randy Sandra'? We would say once. A computer, if carelessly programmed, would say five times. Any passage in quotes must be ignored. Mr Critchley might write: 'As Enoch Powell said . . .' Strike it out before the computer reads it. It doesn't matter what Enoch Powell said. We are only interested in Mr Critchley.

Can it really be true that all the tens of millions of people who can write correct grammatical prose have a unique micro-style? It is a fascinating speculation, but in most cases the question is unimportant. Those angry Government whips on the trail of Mr Critchley probably knew there was only a small number of possible suspects. There were only three or four historians in the ancient world from whom Josephus could have stolen his battle scenes. And anyone wanting to identify the victim of Shakespeare's literary theft need only computerise the writings of half-a-dozen Elizabethan playwrights.

Stylometrics has already served the interests of justice as well as those of scholarship. Dr Morton, in his book *Literary Detection,* * tells how a court defendant was acquitted when it was shown that a statement attributed to him must have been written by somebody else. Successful literary forgeries of any kind are immeasurably more difficult now that people's subconscious habits can be cheaply and quickly subjected to electronic analysis.

*Literary Detection, *Dr Morton (Bowker)*

Measure for measure

King Pellinore was three miles away with a great host.

Malory, *Le Morte d'Arthur*

IF this observation was accurate, how far away was King Pellinore? There is no way to say: for we do not know if the writer was using human miles (1,618 yards), Scottish miles (1,976 yards) or Irish miles (2,020 yards).

Even if this were known, we would still be no nearer to locating the hostile king with precision. For Malory does not reveal whether he means 'yards' in the Roman sense, a pace of a marching soldier, or in the English sense, the length of a man's extended arm.

Had he been writing in a later century, we would have been still further confused. For it might have meant the English statute mile (1,760 yards) or the nautical mile (205 yards).

Whatever the solution to this might have been, it illustrates the veritable jungle of our system of measurements, which the partial introduction of the Metric System has succeeded in making incomparably worse – as I shall explain in a moment.

It is fortunate indeed that historians do not attempt to be too exact about weights and measures, for intolerable confusion would result if they did. Suppose, for example, that we were told that each soldier carried with him a hundredweight of supplies and equipment. Would this mean the English hundredweight, which is 112 pounds, or the American hundredweight, of exactly 100 pounds?

Nor is it always clear what is meant by 'pounds'. A pound can consist of 12, 14, 15 or 16 ounces, 16 being the usual figure. This would be helpful if one was only sure what was meant by an ounce, for at different periods in history the ounce represented different numbers of grains of wheat. Parliamentary statute now recognises both the *avoirdupois* ounce, which weighs 28.35 grams, and the apothecary or 'troy' ounce of 31.1 grams.

All these complicated measurements at least had the advantage of being gradually standardised in the English-speaking world, with tacit agreement to use *avoirdupois* pounds and ounces only, to fix the land mile at 1,760 yards and to eliminate archaic measures like rods, chains, furlongs and leagues.

Then came the Metric System, originally foisted on the world by the ideologues of the French Revolution, which overthrew centuries of progress towards uniformity and which may have done immense harm in retarding scientific education.

People were seduced into accepting the metric International System of Units by the belief that they could grasp it if they could master six basic measurements, the metre, the gram, the ampere, the kelvin, the candela and the mole.

But lurking round the corner were more units, derived from the first group, such as the volt, the joule and the pascal. And these, in turn, often come with prefixes attached to them, like the centi, the milli, the kilo, the mega and the ato. Why, if the system is so simple, do we need all these sub-groups and prefixes?

Or perhaps I am being too conservative. Why not go completely metric? Away with all this nonsense about years and tons of coal equivalent! Why not pass laws allowing people to vote at the age of 567.65 megaseconds? After all, we're an advanced country; we consume energy in the range of exajoules, not mere petajoules like some banana republic. What else should one expect from a nation with a gross national product approaching the teradollar range?

A proposal for a third system of measurements, of a character quite different from the other two, has been proposed by a scientist, Professor David F. Bartlett, of the University of Colorado. At first sight, it will sound even madder. But it does have the virtue of great simplicity, and if the human species is to become a space-faring race, it may be the only one whose measurements prove constant in all circumstances.

Units of length, says Professor Bartlett, should be expressed as fractions of a light-second based on the speed of light through a vacuum of 186,000 miles a second. With this system, Malory would have reported the distance of King Pellinore as being .000016 light-seconds, which admittedly sounds rather more frightening than a comfortable three miles.

Crazy as Professor Bartlett's system may seem, it does have one great advantage. Light-seconds never change, whereas in a fast-moving spaceship, in Einstein's special theory of relativity, conventional measurements are shortened by the speed of the ship. But the speed of light is a fundamental constant, the same to all observers however fast they are moving.

In a sufficiently advanced technological society, Professor Bartlett's system would be the only one that would work. To some extent, it is in use already. We do not measure the distances between the stars in trillions of kilometres – but in light-years.

Back to an unscrupulous ancestor

Knowledge is power.

Francis Bacon

MANKIND's urge to better himself by means of science is a relatively recent phenomenon in world history, becoming dominant only within the last 400 years. The origins of modern science can largely be traced to the writings of the seventeenth-century philosophical genius, Sir Francis Bacon.

To the contemporary mind, which takes it for granted that the proper functions of science are to discover truth and improve material conditions, it is astonishing to realise that Bacon, when he launched his vast intellectual revolution, had to contend against the belief by the educated public that such aims were ridiculous or even irreverent.

It was generally held that scientific knowledge rested, and ought to rest, on the authority of ancient scholars. Aristotle in particular had pontificated on many subjects and until the age of Bacon, people tended to believe that the surest way to settle a scientific question was to look up what Aristotle had said on the matter and quote it as revealed truth.

But this was often dangerous. 'By modern standards,' writes Sir Peter Medawar, 'Aristotle didn't have a clue. His biological writings are a weird mixture of tall stories, credulities and inaccuracies that could have been cleared up by the most elementary observations.'

This patriarch of knowledge stated, for example, as authoritative fact, that people were infertile between the age of 14 and 21. After 21, they could bear children, but the younger the parents the smaller and less healthy were their offspring.

The heart, not the brain, he declared to be the source of human thought: there were five elements, earth, air, fire, water and ether; and the universe was rigidly bounded by 54 heavenly spheres.

It is unfair to blame a scientist for honest mistakes, but Aristotle, and the many philosophers who followed him, committed a much deeper sin: they disdained to check facts. To them debate and reasoning were everything, while experiment was a mere vulgar waste of time. They proclaimed practical inventions to be worthless and in doing so they sabotaged for more than a thousand years the beginnings of Western science.

Bacon's majestic work *The New Organon* (meaning 'the new tool' or 'the new instrument'), published in 1620, was the counterblast to Aristotle, and it succeeded in changing for ever the way in which people looked upon science. Saying 'I have taken all knowledge to be my province,' Bacon looked to a remote future in which technology, hitherto despised, would vastly increase wealth and happiness. He saw through time, perhaps more clearly than anyone before or since.

Bacon himself was no scientist, and of such scientific knowledge that then existed he was exceedingly ignorant. He rejected the heliocentric theory of Copernicus, and he does not seem to have heard of the astronomical discoveries of Kepler.

But these failings are irrelevant. It would be as idle to seek scientific facts in his writings as it would be to discover the geography of Soho from a world atlas. What he did was to create the modern scientific spirit, to insist on the marriage of experiment and theory, bidding people to leave, without regret, the desert that lay behind them, and enter with joy the empire of technological marvels that lay ahead.

What makes the story of Francis Bacon even more interesting is that in political life he was a crook. Observing that 'all rising to great place is by a winding stair' he was unscrupulous, even beyond the standards of his day, in seeking power and riches.

To please Elizabeth I, he laboured to send his best friend the

Earl of Essex to the scaffold. He took a leading role in the judicial murder of Sir Walter Raleigh. Attaining high office under James I, he supervised the illegal torture of an innocent clergyman. But the most spectacular of his crimes was his large-scale acceptance of bribes while Lord Chancellor of England. He awarded cases, not to the highest bidder, but the highest payer. He would accept secret payment from all parties – and then judge the case as he pleased.

But his criminal record does not matter now. His victims have long been in their graves while his philosophy lives on. Shakespeare in his youth assaulted gamekeepers who caught him poaching. But we enjoy his plays without troubling to think of their broken bones. So it is with Bacon. A squalid figure he may have been in political history, but he truly deserves to be called the father of modern science.

INDEX

196